D0295296

WAVES IN PHYSICAL SYSTEMS

WAVES IN PHYSICAL SYSTEMS

CHARLES F. SQUIRE, Ph.D.

PRENTICE-HALL, INC.

Englewood Cliffs, New Jersey

Prentice-Hall International, Inc., *London*
Prentice-Hall of Australia, Pty. Ltd., *Sydney*
Prentice-Hall of Canada, Ltd., *Toronto*
Prentice-Hall of India Private Limited, *New Delhi*
Prentice-Hall of Japan, Inc., *Tokyo*

Current printing (last digit)
10 9 8 7 6 5 4 3
13-946087-x

Library of Congress Catalog Card Number 77-161459
Printed in the United States of America

To WILLIAM V. HOUSTON

Former President, The American Physical Society
Former President, The William Marsh Rice University
Research Physicist, Author, and Friend to Scholars

PREFACE

Our aim in this text has been to give a basic account of wave mechanics in classical and quantum mechanisms at the upper undergraduate level. The book has been written for majors in physics, chemistry, mathematics, geophysics, and engineering. The concepts of impedance, admittance, and response function are used in linear systems. Scalar waves treated for vibrating strings, membranes, and pressure fluctuations give the reader a ready grasp of the mathematical treatment on easily visualized systems. Complex variables, Fourier series, and orthogonal functions are used from the beginning in order to emphasize the methods of quantum mechanics. The course was given in one semester, met three times per week, and at Rice University the students had the option of several laboratory experiments on wave propagation in acoustic and electric transmission lines as well as other topics in this text. A textbook on advanced calculus, normally studied by the reader, served as an essential supplement to the mathematical theory, e.g., properties of Bessel functions. Problem working gives the reader confidence in his knowledge. The blending of modern physics with the classical, together with certain useful aspects of the book, kept student interest and enthusiasm very high. Each topic has been written out in detail rather than touching on numerous subjects. The author has acknowledged herein only a fraction of the scientists who contributed so much to wave mechanics. To students and faculty colleagues, my sincere thanks for help in the making of this small text.

CONTENTS

Chapter 1

OSCILLATING SYSTEMS

In beginning physics courses of mechanics and electricity the harmonic oscillations of a mass on the end of a spring, a torsion pendulum of a mass hung by a fine wire, or an electrical circuit made up of an inductance and a capacitance in series are all treated quite simply and these systems furnish an excellent starting point for a book on wave mechanics. Indeed, the earliest observations in our experience with waves on the surface of water (or on a flexible rope) are that the particles of water (or segment of rope) go up and down as the wave passes by a given point. Thus a discussion of a mechanical oscillator provides an overlap with the familiar and allows mathematical concepts using the complex variable to enter at an early stage. Furthermore, the concepts of impedance and admittance may be developed as useful tools in an understanding of things to come in wave mechanics.

1.1 THE SIMPLE HARMONIC OSCILLATOR

Figure 1.1 shows the simple harmonic oscillator consisting of a mass, m, on the end of a spring whose extention and compression is so small that the restoring force is *linear* and is given by $F = -Ky$. Let us place the oscillator in a vacuum so there will be no viscous drag or added mass on its motion. The diagram shown in Fig. 1.1 (just under the schematic drawing of the oscillator) gives the restoring force law. The other diagrams indicate what an observer would plot for amplitude, y, as time on the stop watch went by, providing he started the watch as the oscillator went through zero on its way upward (observer P) or at the top of its motion (observer Q). Observer P would examine his plot of y vs. t and claim the oscillator obeyed: $y = A \sin (2\pi t/T)$. For observer P, the amplitude of oscillation was zero and moving upward each interval of time, T, called the *period*. On the other

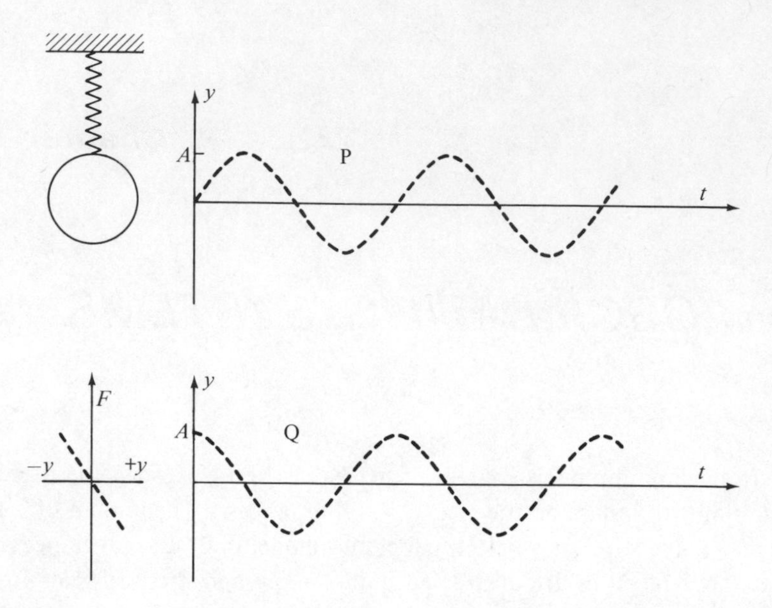

Figure 1.1 Simple Harmonic Oscillator, linear force law, and plot of amplitude as a function of time.

hand, observer Q would claim a behavior characterized by $y = A \cos(2\pi t/T)$ and he would agree that his value of the arbitrary maximum deflection, A, and the value of the period, T, were identical to those reported by P. A third observer on the same oscillator could have started his stop watch at any arbitrary time and could have characterized the motion by $y = A \cos(2\pi t/T + \phi)$. He would then inform P that a value $\phi = \pi/2$ would make then agree and he would inform Q that a value $\phi = 0$ would put them in accord and he would go on to point out that the maximum amplitude, A, and the phase angle, ϕ, are two arbitrary constants associated with this type of mechanical motion. We shall now examine the theoretical basis for these experimental facts. We assume all the mass is at the point $y = 0$ and the spring has no mass.

Applying the laws of mechanics to the mass of the oscillator allows us to state that

$$F = m\left(\frac{d^2y}{dt^2}\right) \equiv m\ddot{y}$$

and that

$$F = -Ky$$

so that

$$m\ddot{y} = -Ky$$

or $$m\ddot{y} + Ky = 0 \qquad \text{Equation of Motion} \tag{1.1}$$

We make use of the property of this homogeneous linear differential equation with constant coefficients by selecting a particular solution $y = Ce^{\alpha t}$ where both C and α may be complex numbers. The solution exists provided the proper value of α is selected. Differentiating twice with respect to time and substituting into Eq. 1.1 gives

$$m\alpha^2 \, Ce^{\alpha t} + K\, Ce^{\alpha t} = 0$$

or $$\alpha^2 + \frac{K}{m} = 0 \qquad \text{with roots } \alpha = \pm i\sqrt{\frac{K}{m}} \qquad \text{pure imaginary number}$$

The general solution is therefore the sum of particular solutions

$$y = C_1 e^{(i\sqrt{K/m})t} + C_2 e^{(-i\sqrt{K/m})t} \equiv C_1[e^{i\omega_0 t} + e^{-i\omega_0 t}] \qquad \text{for the case } C_1 = C_2$$

This special case holds for zero initial velocity at $t = 0$ and where the angular frequency, $\omega_0 = \sqrt{K/m}$, is determined by the mass and spring constant of the oscillator. We keep both terms in this last expression in order to conform with subsequent topics. We take the real part of the complex quantity equal to $y(t)$.

Now

$$\cos(\omega t) = \frac{e^{i\omega t} + e^{-i\omega t}}{2}$$

so that the amplitude of oscillation may be described by

$$y = 2C \cos(\omega_0 t)$$

Obviously, what we have here agrees with observation by Q if $2C = A$ and if $\omega_0 = 2\pi/T$. But what about the osbervations of P and the third party we mentioned? Where is the arbitrary phase constant ϕ? The answer is that the constant C is a complex number such that $C_1 = C_0 e^{i\phi}$ and $C_2 = C_0 e^{-i\phi}$ and the general solution is

$$y = 2C_0 \cos(\omega_0 t + \phi) \equiv A \cos(\omega_0 t + \phi) \tag{1.2}$$

We have again established that our second-order linear differential equation has two arbitrary constants in the solution; i.e., the amplitude $2C_0$ of the oscillator and the phase angle ϕ. Thus the theory matches the experimental observations and is useful because, for example, we could calculate the displacement, y, at any later time, t.

1.2 ANHARMONIC TERM TO OSCILLATOR AS PERTURBATION

It may well be claimed that the above theory is a bit sterile because any real oscillator probably has a spring force given by the linear term already used plus a small quadratic term. Thus we write $F = -Ky + gy^2$ with the magnitude of $g \ll K$. The plot of force vs. displacement in Fig. 1.1 is thus a little steeper for negative values of y and not quite as steep for positive values of y. The equation of motion (without any viscous drag on the system) is now

$$m\ddot{y} + Ky - gy^2 = 0 \tag{1.3}$$

We have seen that Eq. 1.2 was the solution where we neglected the new term, gy^2, so we try a solution

$$y = A \cos(\omega_0 t + \phi) + \xi$$

and we expect the added term, ξ, to be a small quantity such that products like $g\xi$ and $g\xi^2$ may be even smaller and can be neglected. We expect ξ to be some function of time which may be determined. Differentiation of the above equation with respect to time, followed by substitution back into Eq. 1.3, and neglecting small terms gives an equation involving ξ

$$m\ddot{\xi} + K\xi - gA^2 \cos^2(\omega_0 t + \phi) = 0 \tag{1.4}$$

Now we make use of the identity

$$\cos 2x = 2\cos^2 x - 1$$

Substitution gives:

$$m\ddot{\xi} + K\xi - \frac{gA^2}{2}[1 + \cos 2(\omega_0 t + \phi)] = 0 \tag{1.4'}$$

We get a particular solution from the simplified equation

$$m\ddot{\xi}_1 + K\xi_1 - \frac{gA^2}{2} = 0$$

The particular solution is

$$\xi_1 = \frac{gA^2}{2K}$$

We get another particular solution from the equation

$$m\ddot{\xi}_2 + K\xi_2 - \frac{gA^2}{2}\cos 2(\omega_0 t + \phi) = 0$$

to be

$$\xi_2 = -\frac{gA^2}{6K}\cos 2(\omega_0 t + \phi)$$

where we have used the fact that $\omega_0^2 = K/m$. The general solution is thus

$$y = A\cos(\omega_0 t + \phi) + \frac{gA^2}{2K} - \frac{gA^2}{6K}\cos 2(\omega_0 t + \phi) \tag{1.5}$$

The anharmonic term has caused a small positive axis shift of the center about which the mass oscillates (an expansion in solid state physics terminology with larger amplitude A) and the presence of a second harmonic term, $2\omega_0$. Here again the theory gives us deeper insight into the mechanism whereby a medium such as a solid or a liquid may have its atoms driven at a large amplitude with frequency, ω, and because of anharmonic terms the medium itself produces the upper harmonic at 2ω. This effect actually happens with pressure waves sent through a liquid.

1.3 THE DAMPING TERM AS A PERTURBATION OF AN OSCILLATOR

The oscillator used in Fig. 1.1 and in the theory discussed could be placed in a viscous medium and an additional term added to the equation of motion. Actually, if the mass shown in Fig. 1.1 were to move up and down with a very short period, the sphere would act like an acoustic dipole source and the energy intensity of pressure waves produced along the axis of motion would depend upon frequency to the fourth power. The damping term added to the equation of motion would thus depend upon the angular frequency, ω, and would be an interesting problem. In order to take a more convincing damping term in the equation of motion of an oscillator, we consider now the torsion pendulum shown in Fig. 1.2*a*. The diagram shows a sphere of moment of inertia, I, hanging on a rod with a mirror, M, and a fine wire supports the device from a rigid platform above. The twisting or torsion motion is observed by following the spot of light on the ground-glass scale shown in the figure. The torsion pendulum is suspended in a gas in order to give a damping term proportional to the angular velocity. But the damping will be considered as a small perturbation in what follows. The observer may plot what appears in the graph of Fig. 1.2 as the angle of twist, θ, against the time,

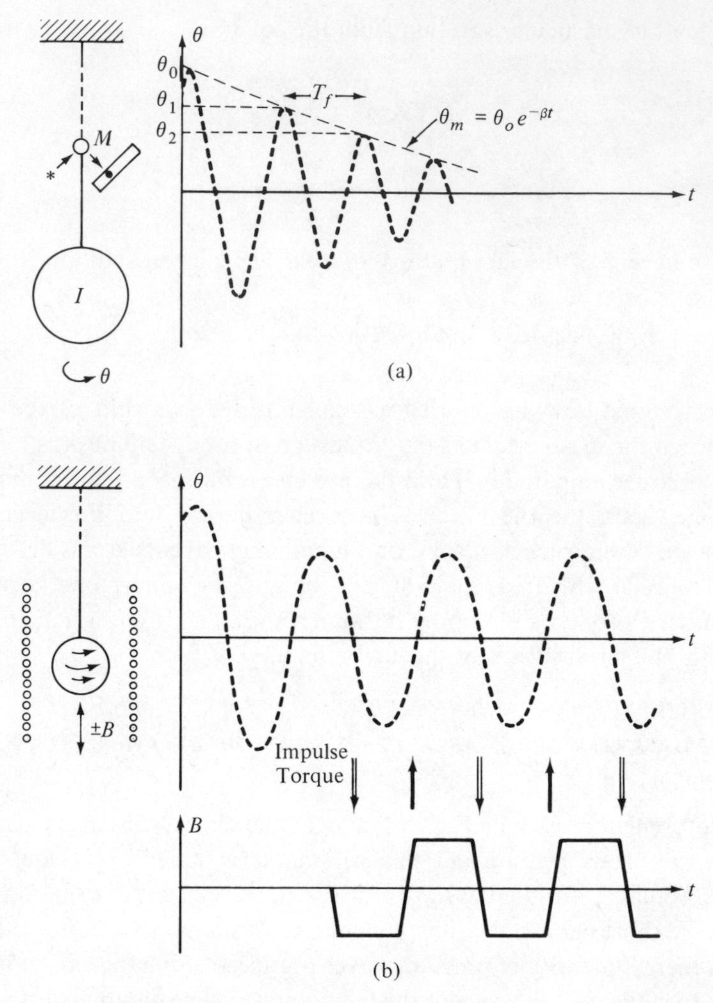

Figure 1.2a Torsion Pendulum, angular displacement as a function of time. **b** Superconducting Torsion Pendulum, angular displacement as a function of time showing steady state amplitude reached after a number of equal impulse torques. The external magnetic field is switched from up to down direction each half period.

t. Here the observer's stop watch was started at an arbitrary point of motion and the time interval for the oscillator to return to a positive maximum could be called an effective period, T_f, as shown. The successive amplitudes become smaller and a careful recording of these would show that the natural logarithm of the ratio of successive amplitudes $\ln(\theta_1/\theta_2) = +\delta$ a constant. In fact the clever observer plots the amplitude $\theta_{\max}$ vs. time, t, on semilog paper and obtains a straight line. We next examine the theoretical basis for these experimental facts.

The equation of motion including the viscous damping action and the linear restoring torque of the torsion fiber is

$$I\ddot{\theta} + R\dot{\theta} + K\theta = 0 \tag{1.6}$$

As before, we seek a solution: $\theta = Ce^{\alpha t}$. The differentiation with respect to time and substitution in Eq. 1.6 gives the equation which the parameter α must satisfy

$$I\alpha^2 + R\alpha + K = 0$$

The roots are

$$\alpha = -\frac{R}{2I} \pm \sqrt{\frac{R^2}{4I^2} - \frac{K}{I}}$$

$$= -\frac{R}{2I} \pm i\sqrt{\frac{K}{I} - \frac{R^2}{4I^2}} \qquad \text{a complex number for } R^2 < 4KI$$

Without damping the natural angular frequency is $\omega_0 = \sqrt{K/I}$. but the effective angular frequency is

$$\omega_f = \sqrt{\frac{K}{I} - \frac{R^2}{4I^2}}$$

The effective period is

$$T_f = \frac{2\pi}{\omega_f}$$

The experimenter can measure the period with very great accuracy; e.g., for a typical period of 30 seconds he can time 100 periods to be 3000 ± 0.2 seconds. Thus he can easily determine the difference between ω and ω_f caused by the damping term. The complex amplitude $C = C_0 e^{i\phi}$ can again be used to write the solution (with $2C_0 = \theta_0$)

$$\theta = \theta_0 e^{-(R/2I)t}\left[\frac{e^{i(\omega_f t+\phi)} + e^{-i(\omega_f t+\phi)}}{2}\right]$$

or

$$\theta = \theta_0 e^{-\beta t} \cos(\omega_f t + \phi) \tag{1.7}$$

where β, the attenuation coefficient, is clearly equal to $R/2I$ in the expression used. Thus, in a time interval $t_2 - t_1 = T_f$, the maximum amplitude will have changed by an amount

$$\Delta\theta = \theta_1 - \theta_2 = \theta_1(1 - e^{-\beta T_f})$$

and taking natural logarithms we have an expression in agreement with experimental observations: $\ln(\theta_1/\theta_2) = +\delta = +\beta T_f$, the logarithmic decrement. All of this is very useful and we give an example of how the theory of the torsion pendulum was used to determine an important quantity, the gyromagnetic ratio of superconducting electrons!

Consider the sphere in Fig. 1.2*b* to be a solid superconductor immersed in liquid helium. Furthermore, let us place a solenoid around the torsion pendulum such that a vertical uniform magnetic field may be switched from the field up to the field down position as desired. It is an experimental fact that the induction $B = 0$ inside the solid superconducting sphere, and this is brought about by supercurrent electrons flowing around the surface of the sphere in order to just eliminate inside the sphere the external field of the solenoid. Such orbital motion of the electrons should have a gyro-magnetic ratio equal to $2m/e$. Viewed from outside the sphere, the supercurrent going around produces an apparent magnetic moment $M = 2\pi R^3 B/\mu_0$ where R is the radius of the sphere, B is the applied field, and μ_0 is the permeability $4\pi \times 10^{-7}$ henry/meter used in MKS units. If we switch the field in a short time Δt from an up position to a down position, i.e., $+B$ to $-B$, then a change in magnetic moment is

$$\Delta M = \frac{4\pi R^3 B}{\mu_0}$$

Such a change in magnetic moment must produce a change in angular momentum. Thus an impulse of torque, $\tau\Delta t$, produces a gain in angular momentum

$$\Delta P_g = \frac{2m}{e} \cdot \frac{4\pi R^3 B}{\mu_0} = \tau\Delta t$$

Conservation of angular momentum requires that the torsion pendulum receive a torque during the brief time, Δt, required to switch the field from $+B$ to $-B$. A changing magnetic field with time produces a circular electric field which pushes on the positive ions. In the experiment the switching is done automatically each time the torsion pendulum swings through $\theta = 0$ and in such a way as to add angular momentum to the system. The pendulum system we have just studied loses angular momentum due to the damping of the viscous medium and we can see that this loss could just be made up each half-period by the driving impulse described above. The situation would then be a steady amplitude of oscillation, θ_s, for the torsion pendulum. The angular momentum lost per half-period is equal to the momentum gained in this steady-state condition. From Eq. 1.7, we have during the free swinging time interval

$$\theta = \theta_s e^{-\beta t} \cos \omega t$$

with $\omega = \omega_f$ in this discussion. The angular momentum is

$$P = I\dot{\theta} = -I\theta_s(\omega e^{-\beta t} \sin \omega t - \beta e^{-\beta t} \cos \omega t)$$

In a time interval $\Delta t = \pi/\omega$ corresponding to one half-period, the loss in angular momentum is

$$\Delta P_l = -I\theta_s \beta\pi + \text{small terms}$$

Letting $\Delta P_l = \Delta P_g$ gives the expression

$$\theta_s = \frac{2m}{e} \frac{4\pi R^3 B}{\mu_0} \frac{1}{I\beta\pi}$$

The experimental measurement of θ_s under known conditions of the parameters R, B, I, and β thus allows the direct measurement of the gyromagnetic ratio $2m/e$ and in the case of the superconductor this is the correct value for orbiting electrons. Much earlier in the present century, Einstein and de Haas performed this experiment for a ferromagnetic sphere and found the gyromagnetic ratio m/e. With barges in the canals in Holland shaking the buildings as they do, it is a wonder de Haas could measure with accuracy! Of course, it may be that the reader believes firmly that the gyromagnetic ratio is $2m/e$ for orbital electrons and that the above superconducting torsion pendulum experiment just confirmed F. London's[1] theory that the magnetic moment of the sphere is indeed $2\pi R^3 B/\mu_0$.

1.4 IMPULSE DRIVING OF AN OSCILLATOR: Response Function

In the above example we spoke of impulse driving the damped oscillator at intervals of time equal to half the period—say, every 15 seconds and for a duration Δt of 10^{-2} second or less. The steady-state amplitude, θ_s, could have been approached from large or from small values of θ and required many hours to achieve. If the torsion pendulum were initially at rest, then both θ and $\dot{\theta}$ would be zero until the time, t_1, of the first impulse. In order to keep things simple, let us neglect viscous damping for the very low angular

[1]F. London, *Physica* 3, 458 (1936).

velocity which the system now acquires. The angular momentum after the first impulse is

$$I\dot{\theta} = \tau\Delta t \cos \omega(t - t_1) \qquad \text{for } t_1 \leq t \tag{1.8}$$

and the equation for the angular amplitude is obtained by integrating with respect to time

$$\theta = \tau\Delta t\left[\frac{\sin \omega(t - t_1)}{I\omega}\right] \qquad \text{for } t_1 \leq t \tag{1.9}$$

Still neglecting friction, the next impulse of the same magnitude occurs at time t_2, which was in our example 15 seconds after t_1, and the linear system allows superposition of the second impulse to give a new equation

$$\theta = \frac{\tau\Delta t}{I\omega}[\sin \omega(t - t_1) + \sin \omega(t - t_2)] \qquad \text{for } t_2 \leq t$$

This process may continue such that the sum may be replaced by an integral equation with a so-called "Green's Function" which is $\left[\frac{\sin \omega(t - t')}{I\omega}\right]$:

$$\theta(t) = \int_0^t \tau(t')\left[\frac{\sin \omega(t - t')}{I\omega}\right] dt' \tag{1.10}$$

According to Eq. 1.9, there is a one-to-one correspondence between the angular displacement $\theta(t)$ and the impulse $\tau\Delta t$. In this example the connecting link is

$$\tilde{\chi}(t - t') = \frac{\sin \omega(t - t')}{I\omega} = \frac{\theta(t)}{\tau\Delta t}$$

Equation 1.10 can then be written in the new symbols

$$\theta(t) = \int_0^t \tau(t')\,\tilde{\chi}(t - t')\,dt' \tag{1.11}$$

The value of this expression is that it is sometimes the starting point of theoretical papers about simple oscillators such as an atom in its own liquid.[2] Also this expression is the starting point in discussions of linear response theory[3] and the quantity $\tilde{\chi}(t - t')$ is called the response function of $\theta(t)$ to τ because it represents the effect of the short duration impulse disturbance of τ at the time t'. The effect in the quantity θ follows the cause at a later time t. In order better to understand the response function, we turn next to the damped oscillator system and discuss its response to a periodic driving force,

[2]Paul C. Martin and Sidney Yip, *Physical Review* **170**, 151 (1968).

[3]R. Kubo, "Linear Response Theory of Irreversible Processes," a chapter in *Proceedings of the International Symposium* held at Aachen, Germany; edited by J. Meixner, North-Holland Publishing Co., Amsterdam (1965).

$F_0 e^{-i\omega t}$. The angular frequency ω of the force may be any value and not necessarily the value ω_f which characterizes the damped oscillator.

1.5 PERIODIC DRIVING FORCE ON AN OSCILLATOR: Complex Susceptibility

The experimental observations on a driven oscillator are presented, Fig. 1.3, and the challenge is to make the theory fit the facts. If the driving

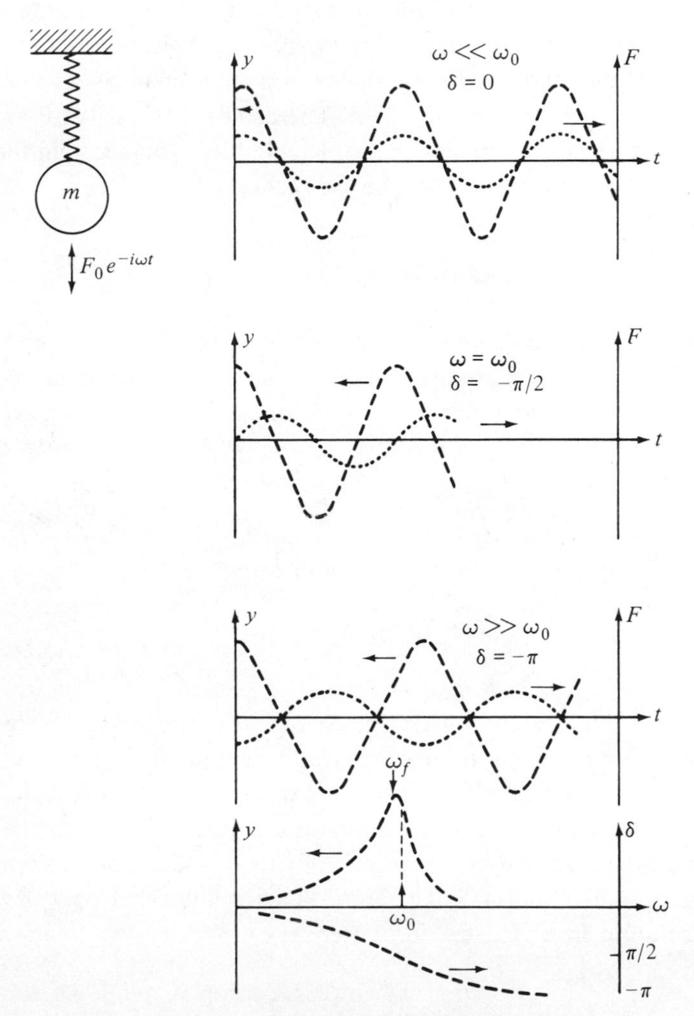

Figure 1.3 Driven Oscillator, amplitude as a function of time, and force as a function of time. Plots are at angular frequencies less than, equal to, and greater than resonance frequency. The fourth plot shows amplitude as a function of angular frequency and also the phase angle between force and amplitude.

force has been on for some time, a steady state of oscillation is observed and Fig. 1.3 shows a segment of time in which the force applied is plotted as well as the amplitude of the driven oscillator. The observer sees that the oscillator amplitude follows the force amplitude at $\omega \ll \omega_0$ in time; i.e., the phase angle $\delta = 0$. The oscillator amplitude under these conditions is small. As the force frequency, ω, increases, the amplitude of the oscillator increases and becomes a maximum at $\omega = \omega_f$ which is very close to, but smaller than, ω_0. When $\omega = \omega_0$ the amplitude of the oscillator is still very large and the phase angle between displacement and force is $\delta = -\pi/2$; i.e., the phase of the oscillation lags behind the driving force. As shown in the fourth plot of Fig. 1.3, the amplitude of the oscillator becomes small again at $\omega \ll \omega_0$ and the phase angle becomes $\delta = -\pi$. The behavior of oscillator amplitude and of phase angle are plotted against driving force frequency in the fourth plot of Fig. 1.3. Let us now examine the theory.

The equation of motion is:

$$m\ddot{y} + R\dot{y} + Ky = [F_0 e^{-i\omega t}] \leftarrow \qquad \text{Take real part} \tag{1.12}$$

if we may assume a damping force simply proportional to the velocity. Not many real systems are given such a simple damping term. We could be less concerned about this if we were discussing the analogous electric circuit

$$L\ddot{q} + R\dot{q} + q/C = [V_0 e^{-i\omega t}] \leftarrow \qquad \text{Take real part}$$

Let us write Eq. 1.12 in a more convenient form

$$\ddot{y} + 2\beta\dot{y} + \omega_0^2 y = ae^{-i\omega t} \tag{1.13}$$

where $\beta = R/2m$, $\omega_0^2 = K/m$, $a = F_0/m$.

The particular solution with zero force was discussed following Eq.1.6. Any oscillation involving the natural frequency, ω_f, of the damped oscillator will vanish with time according to the exponential decay $e^{-\beta t}$. Such motion would be linearly superimposed on the motion caused by the driving force and after steady-state conditions of driven oscillation are established these have usually vanished. Therefore, we are interested only in the particular solution involving the force frequency. We try the solution

$$y = Ce^{-i\omega t} \qquad \text{when } C \text{ is a complex number.}$$

Differentiation with respect to time and substitution in Eq. 1.13 gives

$$-\omega^2 C - 2\beta i\omega C + \omega_0^2 C = a$$

after dividing out common terms.

Solving for the complex amplitude gives

$$C = \frac{a}{(\omega_0^2 - \omega^2) - i(2\beta\omega)} = a\left\{\frac{\omega_0^2 - \omega^2}{(\omega_0^2 - \omega^2) + 4\beta^2\omega^2} + i\frac{2\beta\omega}{(\omega_0^2 - \omega^2)^2 + 4\beta^2\omega^2}\right\} \tag{1.14}$$

Thus the oscillator displacement as a function of time for constant driving force frequency, ω, is the real part of a complex function in ω.

$$y(t) = \mathrm{Re}\left\{F_0 e^{-i\omega t}\left[\frac{\omega_0^2 - \omega^2}{m(\omega_0^2 - \omega^2)^2 + 4m\beta^2\omega^2} + i\frac{2\beta\omega}{m(\omega_0^2 - \omega^2)^2 + 4m\beta^2\omega^2}\right]\right\} \tag{1.15}$$

We may define a complex susceptibility $\chi(\omega) = \chi'(\omega) + i\chi''(\omega)$ and write Eq. 1.15 again, using these symbols as a brief notation

$$y(t) = \mathrm{Re}\{F_0 e^{-i\omega t}[\chi'(\omega) + i\chi''(\omega)]\} \tag{1.15'}$$

In our discussion of the response function for a driven oscillator, Eq. 1.11, the displacement was shown to be (we take the torsion oscillator theory over to the present oscillator)

$$y(t) = \int_0^t \tilde{\chi}(t - t')F_0 e^{-i\omega t'}\, dt' \tag{1.11'}$$

Comparison shows that the complex susceptibility is

$$\chi'(\omega) + i\chi''(\omega) = \int_0^\infty e^{-i\omega t}\tilde{\chi}(t - t')\, dt \tag{1.16}$$

which transforms, by means of an integral, the response function into the susceptibility—a one-sided Fourier transform as it is called. We have called the quantity $\chi(\omega)$ the susceptibility in accord with writers such as Landau and Lifshitz[4] rather than use the term admittance of Kubo[3] because, as we shall see, the admittance is used in discussing the velocity as a function of the force.

[4]L. D. Landau and E. M. Lifshitz, *Statistical Physics*, Pergamon Press (1958), p. 391, London. Second edition, Addison-Wesley Publishing Co., Reading, Mass (1969).

The term susceptibility is sometimes called immittance and also called compliance!

1.6 THE COMPLEX MECHANICAL IMPEDANCE

Returning to Eq. 1.14 for the complex amplitude of the driven oscillator and substituting back the original oscillator symbols

$$C = \frac{a}{(\omega_0^2 - \omega^2) - i2\beta\omega} = \frac{F_0/m}{(K/m) - \omega^2 - i\omega(R/m)} = \frac{F_0}{-i\omega[R - i(m\omega - K/\omega)]} \tag{1.14'}$$

We define the mechanical impedance

$$Z_m = R - i(m\omega - K/\omega) \tag{1.17}$$

The displacement of the driven oscillator is

$$y = \frac{F_0 e^{-i\omega t}}{-i\omega Z_m} \tag{1.18}$$

The velocity is then

$$\dot{y} = \frac{F_0 e^{-i\omega t}}{Z_m} \tag{1.19}$$

The mechanical impedance, Z_m = force/velocity, may be thought as analogous to the electrical impedance which equals the ratio of the voltage to the current. The complex impedance may be set equal to its absolute magnitude multiplied by a phase factor

$$Z_m = |Z_m| e^{-i\phi} \tag{1.20}$$

with

$$|Z_m| = \sqrt{R^2 + (m\omega - K/\omega)^2} \tag{1.21}$$

The velocity is written

$$\dot{y} = \frac{F_0 e^{-i\omega t}}{|Z_m| e^{-i\phi}} = \frac{F_0}{|Z_m|} e^{-i(\omega t - \phi)} \tag{1.22}$$

The velocity phase angle

$$\phi = \tan^{-1}\left(\frac{\omega m - K/\omega}{R}\right) \tag{1.23}$$

and with ω close to zero this angle is close to $-\pi/2$. When $\omega = \omega_0$ the angle $\phi = 0$, and at very large ω the angle ϕ approaches $+\pi/2$. The phase angle, δ, by which the displacement lags the force is simply $\delta = \phi - \pi/2$ and agrees with Fig. 1.3. Of course the amplitude is very large when the impedance is made small in the region of resonance. The admittance is the reciprocal of the impedance, $Y_m = 1/Z_m$, but there is little new in writing $\dot{y} = Y_m F_0 e^{-i\omega t}$ for the driven oscillator.

It is not the general practice to constantly remind the reader that the physical observable, such as $\dot{y}$, is obtained by taking the real parts of the complex expression used above. For example, in the literature[5] discussing an alternating longitudinal stress $\sigma = \sigma_0 \exp(i\omega t)$ to a general linear viscoelastic material, the alternating longitudinal strain is written: $\epsilon = D(i\omega)\sigma$, where $D(i\omega)$ is called the complex longitudinal compliance. Landau (Ref. 4) might have called this last quantity a complex susceptibility. The fact that the real part of the expression on the right-hand side of the above equation is to be taken is understood or implied. We shall follow this convention in the chapters ahead.

PROBLEMS

1. Equation 1.1 in the text is $m\ddot{y} + Ky = 0$ which for convenience may be written $\ddot{y} + \omega_0^2 y = 0$. The solution to be tried in this problem is

$$y = b_1 + C_1 t + b_2 t^2 + C_3 t^3 + b_4 t^4 + \cdots$$

Show that only two arbitrary coefficients, b_1 and C_1, remain and that the power series solution reduces to

$$y = b_1 \cos \omega_0 t + C_1 \sin \omega_0 t$$

Show that the coefficient b_1 is the amplitude at $t = 0$ and that the coefficient C_1 is the velocity at $t = 0$; i.e., that these are the arbitrary initial conditions. Ordinarily these initial conditions are not what is observed and the two constants A and ϕ of Eq. 1.2 are the quantities observed.

2. A damped "ideal" oscillator whose spring has no mass has the following values:

$$m = 200 \text{ grams}$$
$$R = 25 \text{ dynes (sec/cm)}$$
$$K = 5 \times 10^4 \text{ dynes/cm}$$

The oscillator is pushed up a distance $y = 2$ cm and let go at time $t = 0$. How many seconds later will the amplitude be reduced to 0.5 cm?

3. Given a mechanical oscillator with the characteristics listed in Problem 2, what

[5] J. Ross Macdonald, *British Journal Appl. Phys.* **17**, 1347 (1966).

is the magnitude of the mechanical impedance, $|Z_m|$, for this system driven with a periodic force at $\omega = 3$ radians/sec? What is the phase angle between force and velocity?

4. The damped torsion oscillator shown in Fig. 1.2 has a sphere of radius 1 cm and a mass of 100 grams hung from a torsion fiber such that a torque of 10^{-3} dyne cm will cause a twist of $\theta = 10^{-3}$ radian. The delicate oscillator is observed to diminish in angular amplitude by $1/e$ in a time of 3×10^3 seconds. What is the resistive term, R, of this torsion pendulum and what is its effective period T_f?

Chapter 2

WAVES ON A FLEXIBLE STRING: Fourier Series and Fourier Analysis

A study of classical wave mechanics, like the study of oscillating systems discussed in Chapter 1, is best started with a rather simple and idealized mechanical system. The flexible string is such a system and we use it in order to present the basic physical concepts and mathematical tools of a one-dimensional wave propagation. We can understand that a real one-dimensional system, such as a stretched violin string, has stiffness and damping characteristics which would lead the elementary discussion to a more complicated analysis. Let us again begin with a description of what an observer records as the physical facts to be "understood" with a mathematical approach.

Imagine a flexible string stretched between two rigid supports spaced far apart—say, across an entire classroom. Figure 2.1 shows a series of diagrams of the motion of the transverse waves on such a string. The topmost diagram shows a stretched string about to be displaced at a region in its center by an object moving with velocity v_0. At the instant t_1 the object has struck the string and deflected the center as shown. The "cause" has produced an "effect" or response which is shown at a later time, t_2. The two disturbances move outward towards the supports such that their wave crest moves in the x-direction with a velocity c. The disturbances soon hit the end supports and reflection occurs. In a real experiment the subsequent motion becomes quite obscure and the string is seen to vibrate up and down in some manner.

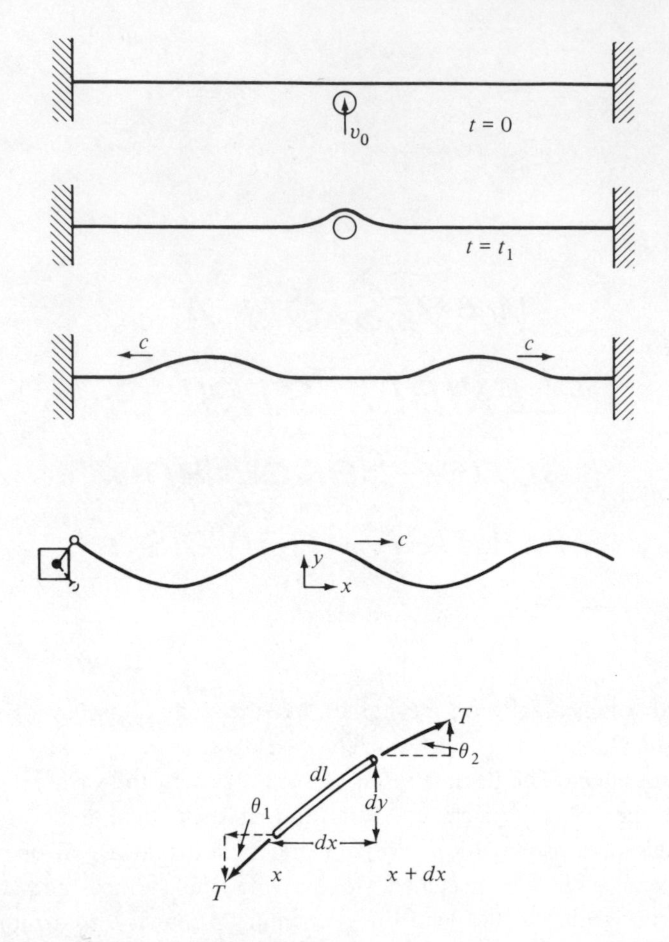

Figure 2.1 Motion on Flexible String, outgoing waves from an oscillator, and diagram of forces on a segment of the string.

The fourth diagram of Fig. 2.1 shows the response of the flexible string to a driving mechanism at the left end, and the wave moving out to the right is meant to extend on as though moving out on an infinite string. The observer again sees the wave maximum moving outward with a characteristic velocity c. In actual practice the outgoing wave would be attenuated in amplitude as it moves out. These physical facts or observations could be enriched by changing the mass per unit length, ρ, of the string and by changing the tension, T, with which the string is stretched between the supports. The experiment would show that the wave velocity is given by $c = \sqrt{T/\rho}$. Figure 2.1 does not include all the wave phenomena possible to observe and in particular we shall wish to discuss so-called standing waves between the rigid supports.

Let us now turn to an analysis of the forces acting on the oscillating system and obtain the wave equation of motion.

In Fig. 2.1 the bottom diagram shows an enlarged segment, *dl*, of the string for purposes of describing the forces on an element of the string. The tension component in the *y*-direction at the right end of the infinitesimal length is different than that at the left end. The net transverse force is the value at $X + dx$ minus that at X

$$dF_y = \left[(T \sin \theta) + \frac{\partial(T \sin \theta)}{\partial x}\, dx\right] - (T \sin \theta)$$

$$dF_y = \frac{\partial(T\, \partial y/\partial x)}{\partial x}\, dx \qquad \text{for small displacement } y \; (\sin \theta = \tan \theta = \partial y/\partial x)$$

$$dF_y = T\frac{\partial^2 y}{\partial x^2}\, dx \tag{2.1}$$

This force must equal the mass times the acceleration

$$T\frac{\partial^2 y}{\partial x^2}\, dx = \rho\, dx \frac{\partial^2 y}{dt^2} \tag{2.2}$$

Thus the equation of motion or wave equation becomes

$$\frac{\partial^2 y}{\partial x^2} = \frac{\rho}{T}\frac{\partial^2 y}{\partial t^2} = \frac{1}{c^2}\frac{\partial^2 y}{\partial t^2} \tag{2.3}$$

with the velocity $c = \sqrt{T/\rho}$ to be identified as such after a solution to Eq. 2.3 has been obtained. We shall take the convention of symbols used by the electrical engineers in particular but also by many modern scientists. The solution which separates the variables is

$$y = Y(x)e^{j\omega t} \tag{2.4}$$

where we must take the real part and where $j = -i$ in this trivial change from the notation in Chapter 1. By differentiation with respect to time we obtain

$$\frac{\partial^2 y}{\partial t^2} = -\omega^2 Y(x) e^{j\omega t} \tag{2.5}$$

and

$$\frac{\partial^2 y}{\partial x^2} = \frac{\partial^2 Y(x)}{\partial x^2} e^{j\omega t} \tag{2.6}$$

The differential equation for the space dependent part must be

$$\frac{\partial^2 Y}{\partial x^2} + \frac{\omega^2}{c^2} Y = 0 \tag{2.7}$$

This type of differential equation was solved in Chapter 1 by use of the trial solution

$$Y = Ce^{\gamma x} \tag{2.8}$$

and the condition is once again similar in that

$$\gamma = \pm i\frac{\omega}{c} = \mp j\frac{\omega}{c} \tag{2.9}$$

In this way we have the complete solution as the sum of the special solutions; i.e., the sum of the wave moving to the right and the wave moving to the left

$$y = [C_+e^{-jkx} + C_-e^{+jkx}]e^{j\omega t} \qquad \text{with } k = \omega/c \tag{2.10}$$

Consider the string clamped at both ends. Suppose now the wave moving to the right were to be perfectly reflected by a rigidly clamped support at a position $x = L$. Then the return wave has the same amplitude as the outgoing wave *but* is shifted in phase by 180° in order to make the value of $y = 0$ at $x = 0$, We may write this as

$$C_+ = C_-e^{j\pi} \tag{2.11}$$

Then our total displacement is written

$$y = [C_+e^{-jkx} + C_+e^{-j\pi}\cdot e^{+jkx}]e^{j\omega t} \tag{2.12}$$

or

$$y = [e^{-jkx} - e^{+jkx}]C_+e^{j\omega t} = -2j \sin kxC_+e^{j\omega t} \tag{2.12$'$}$$

The complex amplitude may be written

$$-2jC_+ = Ae^{j\phi} \tag{2.13}$$

when A is a real number. Then our total displacement becomes:

$$y = A \sin kxe^{j(\omega t + \phi)} \tag{2.14}$$

whose real part is taken.

Now for the wave amplitude to vanish at $x = L$ we demand that $\sin kL = 0$ and so $k = \pi/L$ or any integer multiple; e.g., $n\pi/L$. We thus have a set of allowed modes of motion of frequency $\omega_n = \pi nc/L$:

$$y_n = A_n \sin k_n x e^{j(\omega_n + \phi_n)} \tag{2.15}$$

whose real part is taken. If a number of modes are present at one time on the string, we write (as Fourier series of orthogonal functions)

$$y = \sum_{n=1}^{\infty} A_n \sin\left(\frac{\pi n x}{L}\right) \cos\left(\frac{\pi n c}{L} t - \phi_n\right) \tag{2.16}$$

or

$$y = \sum_{n=1}^{\infty} \sin\left(\frac{\pi n x}{L}\right)\left\{B_n \cos\left(\frac{\pi n c}{L}\right)t + C_n \sin\left(\frac{\pi n c}{L}\right)t\right\} \tag{2.17}$$

where B_n is the amplitude and C_n the velocity amplitude.

Figure 2.2 shows two possible modes on the flexible string stretched between two rigid supports with $n = 1$ the fundamental and $n = 2$ the first

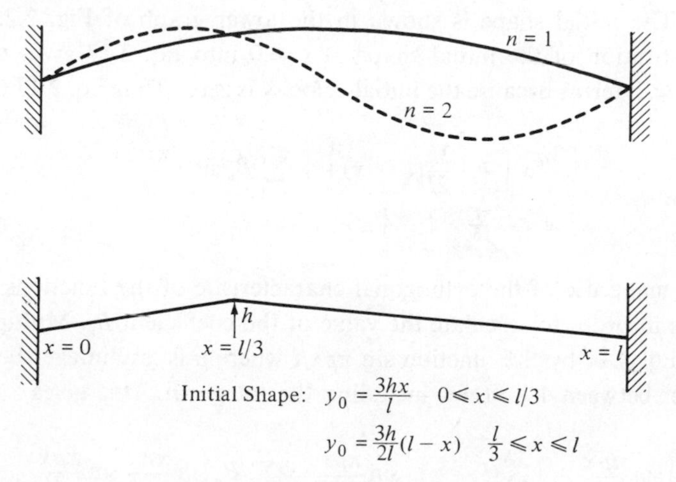

Figure 2.2 Characteristic Modes of Motion on Flexible String and diagram of string pulled aside to give an initial shape at $t = 0$.

harmonic. A simple addition of these two modes gives the resultant amplitude, y, along the distance x and in this example could be a large positive deflection in the left portion of the string and a small negative bulge at the right end. The important thing we have learned from this is that any continuous shape we might have on the string (such as pulling it aside at a point $x = l/3$ down the length of the string) can be closely approximated by the series of Eq. 2.16 or 2.17. We must assign the proper amplitude to any mode used in the build-up of the desired shape and, of course, we may not wish to use some modes at all in the construction task.

What we describe here is called Fourier analysis of the shape of the string by use of a set of known functions—in this case the functions $\sin(\pi nx/L)$. The exciting thing about this theoretical approach is that the resulting shape of the string can be predicted for any time, t, later after the string has been released from its initial shape. In order to emphasize the importance of this method of analysis we anticipate discussions in later chapters of this book to say that the wave function used to describe argon atoms in the gaseous state and to describe electrons in the metallic state are represented by a Fourier series such as Eqs. 2.16 and 2.17. When we need to describe the wave function of a two- or three-dimensional system, the set of functions used in the series representation will quite naturally be different from the simple $\sin(\pi nx/L)$ functions used in the present one-dimensional string. In order to bring this theory into sharp focus let us calculate the shape for the example cited of pulling the string aside a small distance, h, at a point $x = l/3$. The initial shape is shown in the lower graph of Fig. 2.2.

Substitution of the initial shape at $t = 0$ into Eq. 2.17 gives only the $B_n \sin(\pi nx/l)$ terms because the initial velocity is zero. Thus Eq. 2.17 becomes

$$\underset{0 \le x \le \frac{l}{3}}{\left[\frac{3h}{l}x\right]} + \underset{\frac{l}{3} \le x \le l}{\left[\frac{3h}{2l}(l-x)\right]} = \sum_{n=1}^{\infty} B_n \sin\left(\frac{\pi nx}{l}\right) \tag{2.18}$$

We now make use of the orthogonal characteristic of the functions used in the series in order to calculate the value of the coefficient B_n. Multiply both sides of Eq. 2.18 by the function $\sin \pi px/l$ where p is any integer belonging to the set between 1 and ∞, including the integer n. This gives

$$\underset{\substack{\text{for } x \text{ from } 0 \text{ to } l/3\\ \text{and zero elsewhere}}}{\left[\frac{3h}{l}x\right]\sin\frac{\pi px}{l}} + \underset{\substack{\text{for } x \text{ from } l/3 \text{ to } l\\ \text{and zero elsewhere}}}{\left[\frac{3h}{2l}(l-x)\right]\sin\frac{\pi px}{l}} = \sum_{n=1}^{\infty} B_n \sin\frac{\pi nx}{l}\sin\frac{\pi px}{l} \tag{2.19}$$

Next we integrate each term from 0 to l and obtain

$$\int_0^{l/3}\left(\frac{3h}{l}x\right)\sin\frac{\pi px}{l}\,dx + \int_{l/3}^{l}\frac{3h}{2l}(l-x)\sin\frac{\pi px}{l} = \frac{l}{2}B_p \tag{2.20}$$

The reason for the above is $\int_0^l \sin(\pi nx/l)\sin(\pi px/l)\,dx$ is zero unless n equals p and in this latter case the integral is just $l/2$. Thus of all the many terms in the series $\sum_{n=1}^{\infty} B_n \sin(\pi nx/l)$ only the one term, $B_p \sin(\pi px/l)$, gives a non-zero value when multiplied by $\sin(\pi px/l)$ and integrated from 0 to l.

The Fourier coefficient is computed from the definite integrals involving the initial shape of the string. We carry this out in detail as shown in the next steps:

$$B_p\frac{l}{2} = \frac{3h}{l}\left(\frac{l}{\pi p}\right)^2\left[\sin\frac{\pi px}{l} - \frac{\pi px}{l}\cos\frac{\pi px}{l}\right]_0^{l/3}$$

$$-\frac{3h}{2l}\cdot l\cdot\frac{l}{\pi p}\left[\cos\left(\frac{\pi px}{l}\right)\right]_{l/3}^{l} - \frac{3h}{2l}\left(\frac{l}{\pi p}\right)^2\left[\sin\left(\frac{\pi px}{l}\right) - \frac{\pi px}{l}\cos\frac{\pi px}{l}\right]_{l/3}^{l} \quad (2.21)$$

Substituting the limits and collecting terms gives

$$B_p = \frac{3h}{\pi^2 p^2}\left[\sin\frac{\pi p}{3} - \sin\pi p\right] \quad (2.22)$$

But $\sin \pi p$ is always zero for integral values of p, and thus there is just the expression

$$B_p = \frac{3h}{\pi^2 p^2}\sin\frac{\pi p}{3} \quad (2.23)$$

The complete expression for the transverse displacement at $t = 0$ and for all later time t becomes

$$y = \sum_{p=1}^{\infty}\frac{3h}{\pi^2 p^2}\sin\frac{\pi p}{3}\sin\frac{\pi px}{l}\cos\frac{\pi pct}{l} \quad (2.24)$$

Evaluation of the first three terms (using the fact that $\sin \pi/3 = \sqrt{3}/2$, $\sin 2\pi/3 = \sqrt{3}/2$, etc.) gives the following:

$$y = \frac{3^{5/2}h}{2\pi^2}\left[\sin\frac{\pi x}{l}\cos\frac{\pi ct}{l} + \frac{1}{4}\sin\frac{2\pi x}{l}\cos\frac{2\pi ct}{l} - \frac{1}{16}\sin\frac{4\pi x}{l}\cos\frac{4\pi ct}{l}\cdots\right] \quad (2.25)$$

Note that the term $p = 3$ is zero and in the example given this is necessary because we need to build up the displacement at $x = l/3$.

The proof of the statement that $\int_0^l (\sin mx \sin nx)\,dx = 0$ unless $m - n = 0$, in which case the integral becomes $= l/2$, may be seen as follows:

$$\sin nx = \frac{e^{inx} - e^{-inx}}{2i} \quad \text{and} \quad \sin mx = \frac{e^{imx} - e^{-imx}}{2i}$$

$$\sin nx \sin mx = -\tfrac{1}{4}[e^{ix(m+n)} - e^{ix(m-n)} - e^{ix(n-m)} + e^{-ix(m+n)}]$$

Everything cancels unless $m - n = 0$, and in this case the above product is just $\frac{1}{2}$.

We emphasize that it is possible to express any function which is continuous and has the boundary conditions of zero amplitude at each end by the Fourier series of the type used in the above theory.

2.1 ONE DIMENSIONAL PARTICLE WAVE: de Broglie Wavelength, Schrödinger Wave Equation

Equation 2.7 for the one-dimensional wave on the flexible string was written

$$\frac{\partial^2 Y}{\partial x^2} + \frac{\omega^2}{c^2} Y = 0 \tag{2.7'}$$

Substituting the wave length $\lambda = c2\pi/\omega$, we obtain

$$\frac{\partial^2 Y}{\partial x^2} + \left(\frac{2\pi}{\lambda}\right)^2 Y = 0 \tag{2.26}$$

The momentum of an elementary particle such as a single helium atom can be represented by the de Broglie wave length, λ, and Planck's constant, h, through the equation

$$p = \frac{h}{\lambda} \tag{2.27}$$

Thus the wave equation for such a particle in a one-dimensional box is

$$\frac{\partial^2 Y}{\partial x^2} + \frac{4\pi^2 p^2}{h^2} Y = 0 \tag{2.28}$$

The momentum for such a free particle is related to the particle's energy, E, by

$$\frac{p^2}{2m} = E \tag{2.29}$$

Thus the wave equation for the one-dimensional particle system becomes the famous Schrödinger wave equation

$$\frac{\partial^2 Y}{\partial x^2} + \frac{8\pi^2 m E}{h^2} Y = 0 \tag{2.30}$$

Using as a solution one of the modes of motion, e.g., the nth mode,

$$Y_n = A_n \sin\left(\frac{\pi n x}{l}\right) \tag{2.31}$$

The characteristic energy, E_n, associated with this characteristic wave function Y_n, may be calculated by differentiation and substitution back into the wave equation. This gives

$$E_n = \frac{n^2h^2}{8ml^2} \tag{2.32}$$

which shows that the energy is given in *quantized* units, For a helium atom in a box of 1 mm length, $E_n = 8.25 \times 10^{-29}n^2$ in energy units of ergs and these energy quanta are very close together if we increase n by one. Furthermore, if we consider the helium atom to have the average kinetic energy of a one-dimensional gas $E = \frac{1}{2}kT$ where k is Boltzmann's constant $= 1.38 \times 10^{-23}$ joule/K and where the absolute temperature, T, is, say, 300 K. The atom has the encrgy of about 3×10^{-14} erg and setting $3 \times 10^{-14} = 8.25 \times 10^{-29}n^2$ we calculate that at about room temperature many of the atoms must use the quantum number $n \approx 10^7$ as an order of magnitude. The quantized nature of the energy or of the momentum for a helium atom at room temperature would be most difficult to observe and many energy "levels" or values are very close together. Statistically we would expect the occupation number of any specific energy value would be a very small fraction indeed. All of these ideas are used in constructing a statistical mechanical model of a gas in one or more dimensions. For example, the so-called "sum over states" for a single particle in one dimension is written

$$Z = \int_0^l \frac{2}{l} \sum_{n=1}^{\infty} e^{-n^2h^2/8ml^2kT} \sin^2\left(\frac{\pi nx}{l}\right) dx \tag{2.33}$$

$$Z = \int_0^{\infty} e^{-n^2(h^2/8ml^2kT)}\, dn \tag{2.34}$$

$$Z = (2\pi mkT)^{1/2}\frac{l}{h} \tag{2.35}$$

We have obviously drawn on an elementary knowledge of classical statistical mechanics in the above expression for the purpose of giving an example of the use of characteristic functions sin $(\pi nx/l)$ and of characteristic energy $n^2h^2/8ml^2$. Note in the above that the amplitude of the wave used was $\sqrt{2/l}$ in order to make $\int_0^l |Y_n|^2\, dx = 1$.

Waves representing single particles in modern physics have been introduced to the young generation of scientists and engineers through well-written and illustrated texts, beginning even at the high school science level and increasing in scientific rigor through introductory courses in college physics and chemistry. The need for a wave character was implied in using the wave Eq. 2.26 and introducing the de Broglie wavelength through Eq. 2.27. Here we assumed that the reader will recall how de Broglie (Thesis, Paris, 1924) was led to the wave theory of matter by considering the Bohr circular orbit of a particle in a central field of force as those orbits for which the angular momentum is equal to an integer times $h/2\pi$. Thus the circumference of the orbit is an integer multiple of the wavelength corresponding to the classical momentum, and in a later chapter we shall make these ideas more plausible. We assume that the reader knows of the need for the wave field in space and time to explain the diffraction and interference experiments done on beams of material particles by Stern and others.

What is the physical meaning of the wave function used for the particle in the above paragraphs? Have we abandoned the idea of the corpuscular nature of elementary particles and now consider the particle spread out over space as in the wave function, Eq. 2.31? Certainly in the use of the wave function $\sin(\pi n x/l)$ for the calculation of E_n and of Z, we did not ask *where* the particle was located in the box at an instant of time. All we required was that the particle *was* in the box somewhere with a certainty of one out of one or 100% We required that,

$$\int_0^l |Y_n|^2 \, dx = 1$$

and so the square of the wave amplitude at a point x is a measure of the probability that the particle is located in that small region of space dx about x. Note that in order to calculate the exact energy, E_n, for the particle having quantum number, n, we gave up an exact knowledge of the location of the particle in the box. Heisenberg (1927) recognized that in quantum wave theory the spatial coordinates of a mass point and the components of its momentum can be defined only within a certain margin such that a reciprocal relation exists between the definition of the position and that of the momentum.

A localized particle can be described by a wave packet made up of monochromatic waves as we have done on the string by a Fourier series and illustrated by Eq. 2.24. The amplitudes of the waves are appreciably different from zero only inside a small finite region of space where the particle is "located." The wave field itself which we have described here is never probed as such. The situation is similar to optics where we do not measure the electric vector, **E**, in the wave field as to magnitude and phase but instead measure the intensity which is proportional to $|E|^2$ at some position x and averaged over considerable time. With particle beams we also measure intensity at some point x by use of suitable devices. The intensity is proportional to the square of the resultant wave amplitude. The resultant wave may be constructed mathematically with superposition of plane monochromatic waves. We have not quite abandoned the classical idea of localizing the particle in space and time but we have certainly endowed the particle with a broader set of properties by used of the wave description. The particle velocity is determined by the velocity of the group of waves, the group velocity, and not by the phase velocity of the individual monchromatic waves making up the wave packet. Later in this book we shall examine other particle waves and at this point we deem it best to continue our discussion of one-dimensional waves.

2.2 ONE DIMENSIONAL WAVES IN A PERIODIC SYSTEM: Brillouin Zones

Of considerable importance to our basic understanding of the characteristic vibrations of atoms in a solid is the way in which the motion of a continuous flexible string is modified when we lump the mass at intervals a and bind the mass points by a massless spring having a force constant, α. Figure 2.3 shows the linear array of atoms as in a solid, or we can look upon

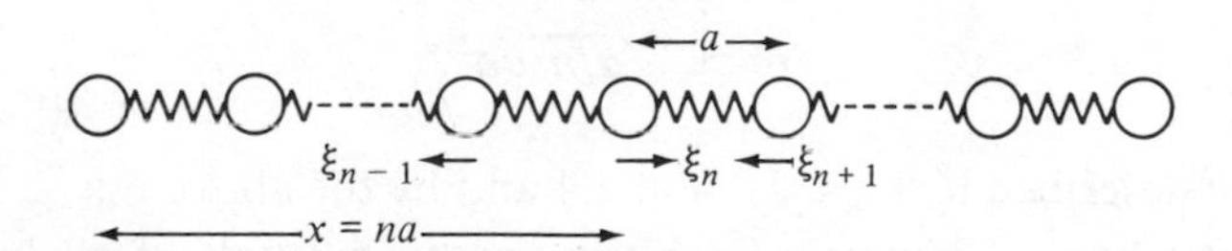

Figure 2.3 Linear Array of Identical Atoms, separated by springs of force constant α and distance a.

the device as a weighted string of very special construction. The particle displacement of the nth atom is denoted by ξ_n and the spring on either side of the nth atom is stretched or compressed according to the relative motion of the $(n - 1)$th atom and the $(n + 1)$th atom. The equation of motion of the nth atom is

$$m\ddot{\xi}_n + \alpha[\xi_{n+1} - \xi_n) - (\xi_n - \xi_{n-1})] = 0$$

or

$$m\ddot{\xi}_n - \alpha(2\xi_n - \xi_{n+1} - \xi_{n-1}) = 0 \tag{2.36}$$

The solution we seek is a wave like motion which is very much like our earlier theory except that the motion is not transverse but is compressional. We try the wave

$$\xi_n = Ae^{-i\omega t} \cdot e^{i(\omega/c)x} = Ae^{-i\omega t} \cdot e^{i\mathbf{q} \cdot \mathbf{n}\mathbf{a}} \tag{2.37}$$

where c is the wave velocity, $\mathbf{q}$ is the wave vector, $\mathbf{q} = \omega/c$, and the distance x to the particle n is equal to na. Differentiation and substitution into Eq. 2.36 gives

$$-m\omega^2 + \alpha(2 - e^{iqa} - e^{-iqa}) = 0$$

or

$$\omega^2 = \frac{2\alpha}{m}(1 - \cos qa) = \frac{4\alpha}{m}\sin^2\left(\frac{qa}{2}\right) \tag{2.38}$$

From this last equation we see that frequencies ($2\pi\nu = \omega$) are determined as shown in Fig. 2.4 where we plot the square root of Eq. 2.38. The sine function runs from a value 0 to a value 1. When $q = \pm\pi/a$ we have the maximum allowed frequency $\omega_0 = 2\sqrt{\alpha/m}$, and when q is very small we have allowed frequencies proportional to the wave number

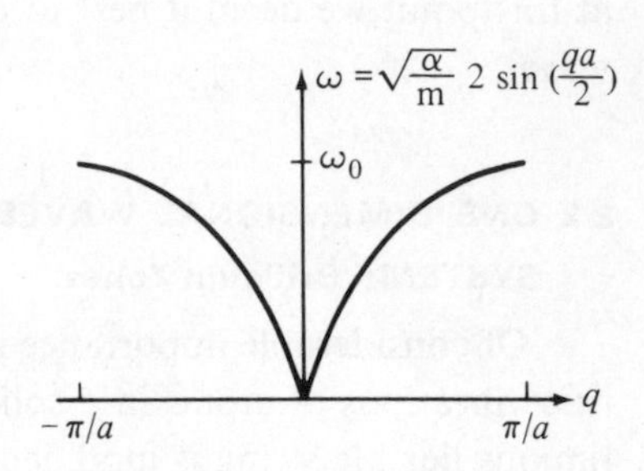

Figure 2.4 Plot of angular frequency ω as a function of wave vector **q**.

$$\omega \simeq \sqrt{\alpha/m}\,qa \tag{2.39}$$

The system described by Figs. 2.3 and 2.4 and by the above Eqs. 2.36 to 2.39 acts as a mechanical low-pass filter with a sharp cutoff of all frequencies above ω_0. The system is formally identical to acoustic and electric filter networks and we shall see in the next chapter a similar behavior for electron waves in a periodic potential. The theory described above for the atoms in a solid stems from the early work of Born[1] and his students, especially M. Blackman. The zone of allowed q values is called the Brillouin[2] zone for the elastic waves. If we tried to propagate a wave of higher frequency than ω_0, the system would refuse the input wave by reflecting it and this can be regarded as Bragg reflection. The mechanical system offers an infinite, purely imaginary impedance to the input wave signal of $\omega > \omega_0$. We shall describe these impedance ideas in Chapter 4. The wave vector, **q**, for elastic waves in a

[1] M. Born and Th. von Karman, *Physik Zeit*, **13**, 297 (1912). M. Blackman, *Proc. Roy. Soc.* (London), **148**, 384, (1935).

[2] L. Brillouin, *Wave Propagation in Periodic Structures*, McGraw-Hill Book Co., New York (1946).

solid will be used again in Chapter 9 where we discuss scattering of X-ray and of neutrons.

The above theory is so important to solid-state theory and to the theory of specific heat of solids that we continue the discussion now to a linear array of diatomic molecules. Figure 2.5 shows the particles of mass m and m'

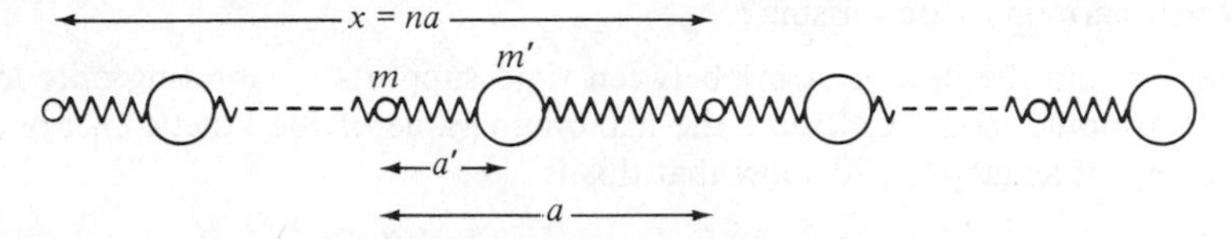

Figure 2.5 Linear Array of Diatomic Molecules.

spaced a distance a and a'. The force constants are α and α' respectively between different molecules and between the atoms of a given molecule. The equation of motion is the pair of equations

$$m\ddot{\xi}_n = -\alpha(\xi_n - \xi'_{n-1}) - \alpha'(\xi'_n - \xi_n)$$

$$m'\ddot{\xi}'_n = -\alpha(\xi'_n - \xi_{n+1}) - \alpha'(\xi'_n - \xi_n) \tag{2.40}$$

We consider solutions to each of the above

$$\xi_n = Ae^{-i\omega t} \cdot e^{i\mathbf{q}\cdot na}$$

$$\xi'_n = A'e^{-i\omega t} \cdot e^{i\mathbf{q}\cdot(na+a')} \tag{2.41}$$

We differentiate, substitute into the proper Eq. 2.40, and from the secular equation formed by the coefficients of A and A' we obtain[3]

$$\omega_{\pm}^2 = \frac{m+m'}{mm'}\left(\frac{\alpha+\alpha'}{2}\right) \pm \sqrt{\left|\left(\frac{m+m'}{mm'}\right)\left(\frac{\alpha+\alpha'}{2}\right)\right|^2 - \frac{4\alpha\alpha'}{mm'}\sin^2 qa} \tag{2.42}$$

The lower branch, ω_-, is called the "acoustic mode" and the upper branch, ω_+, is called the "optical mode." The Brillouin zone boundary occurs at $q = \pm\pi/2a$ and there is now a gap of non-allowed frequencies. The filter action of this mechanical system is like a band-stop filter in electrical networks. When $q = 0$ for the optical branch the two atoms in the molecule vibrate against one another; i.e., we do not have a wave propagated through the system; the diatomic molecules just vibrate independently.

[3]See, for example, M. Blackman, "The Specific Heat of Solids," Vol. VII, Crystal Physics, *Handbuch der Physik*, edited by S. Flugge, Springer-Verlag, Berlin (1955).

PROBLEMS

1. A flexible string is stretched between rigid supports under tension, T, with length, l, and mass per unit length, ρ. The string is pulled aside a small distance, h, at the midpoint and is released at time $t = 0$. Give the complete expression for the displacement, y, with the Fourier series method used in this chapter. Which harmonics are missing?

2. Assume that the flexible string between rigid supports is vibrating only in its nth harmonic mode. Calculate the maximum value of the kinetic energy of a segment of length dx and show that this is

$$dE_n^{\max} = \frac{\omega_n^2 \rho}{2}(B_n^2 + C_n^2) \sin^2\left(\frac{\omega_n x}{c}\right) dx$$

Calculate the maximum kinetic energy of the entire string of length l which the nth harmonic possesses.

3. A flexible string of 12-gram mass and 200-cm length is stretched to a tension of 10^6 dynes. What is the frequency of fundamental vibration? If the displacement amplitude of the fundamental mode is 1 cm at the center of the string, what is its velocity amplitude at this point? What is the total energy in this mode of the entire string?

4. Figure 1.2b in Chapter 1 shows the magnetic field induction, B, plotted as a function of time. This is very nearly a square wave or step function. Jean Baptiste Fourier (1768-1830) was the first to construct the infinite series of simple continuous functions which could represent the step function shown in Fig. 1.2b and caused quite an uproar in those days. Show that $B(t)$ of Fig. 1.2b in Chapter 1 can be represented by a sum of sine functions made up of the fundamental and all odd-numbered harmonics; i.e., 3, 5, 7, etc.

Chapter 3

ELECTRON WAVES IN A METAL: Bloch Waves

The waves on a string, treated in Chapter 2, can now be taken over in a mathematical way to discuss the electron moving in a one-dimensional periodic lattice as in a metal. The wave equation for the string becomes a wave equation for an electron by use of the de Broglie relation: $p = h/\lambda$ where p is the momentum of the electron, h is Planck's constant, and λ is the wavelength. The amplitude of the wave takes on the quantum mechanical meaning that its absolute value squared is a measure of the probability of the electron being at that position in space with a certain characteristic energy. The new wave equation (Schrödinger wave equation for the electron is similar to Eq. 2.30 with the inclusion of a potential energy. We have a wave amplitude, ψ, whose absolute value squared gives the probability for the electron being at X and whose wave equation is

$$\nabla^2\psi + \frac{8\pi^2 m}{h^2}(E - V)\psi = 0 \tag{3.1}$$

If we take the potential energy $V = \alpha \cos 2\pi x/a$, we have the electron moving through a periodic potential of amplitude α and repeating with the lattice spacing a. The wave equation is now

$$\frac{d^2\psi}{dx^2} + \frac{8\pi^2 m}{h^2}\left(E - \alpha \cos \frac{2\pi x}{a}\right)\psi = 0 \tag{3.2}$$

This typical equation is sometimes known as Mathieu's equation. If the potential is zero so that the wave equation represents completely free electrons, the solution of the equation is a simple wave such as

$$\psi = e^{2\pi i \mathbf{k} x} \tag{3.3}$$

where $\mathbf{k}$ is positive or negative and the energy is given by $E = k^2(h^2/2m)$. We may take the amplitude of the wave such that the value

$$\int_0^L |\psi|^2 \, dx = 1 \tag{3.4}$$

We wish to show that the existence of the periodic potential requires a wave[1]

$$\psi = e^{2\pi i \mathbf{k} x} u(x) \tag{3.5}$$

where $u(x)$ is periodic in lattice spacing a.

Let us solve the Mathieu equation by the Fourier series as follows:

$$\psi = e^{2\pi i \mathbf{k} x} + \sum_{n=1}^{\infty} A_n e^{2\pi i (\mathbf{k} - n/a) x} \tag{3.6}$$

where n are positive or negative integers but *not* zero. Just as in the theory for a simple string, we must calculate the Fourier coefficients, A_n, used in the series. The orthogonality of the waves are used to show that

$$A_{+1} = \frac{\alpha m a}{2h^2(k - 1/2a)} \tag{3.7}$$

$$A_{-1} = -\frac{\alpha m a}{2h^2(k + 1/2a)} \tag{3.8}$$

In first-order approximation the electron wave function is

$$\psi = e^{2\pi i \mathbf{k} x}\left[1 + \frac{\alpha m a}{2h^2}\left(\frac{e^{-2\pi i x/a}}{k - 1/2a} - \frac{e^{2\pi i x/a}}{k + 1/2a}\right)\right] \tag{3.9}$$

Obviously something drastic happens to ψ at values of the wave vector $\mathbf{k} = 1/2a$ and this is known as the Brillouin zone edge. Details are now given in the following paragraphs.

We shall write again Eq. 3.5 as

$$\psi = \{A_0 e^{2\pi i \mathbf{k} x} + \sum_{n \neq 0} A_n e^{2\pi i (\mathbf{k} - n/a) x}\} \tag{3.10}$$

Thus ψ is equal to the free wave function plus some other term. We

[1] F. Bloch, *Z. Physik*, **52**, 555 (1928).

will show that, in general, this other term is rather small in most cases. Using the value $E_0 = (h^2/2m)k^2$ and Eq. 3.2, we get

$$\frac{d^2\psi}{dx^2} + \left[\frac{8\pi^2 m}{h^2}\left(E - E_0 - \alpha \cos\frac{2\pi x}{a}\right) + 4\pi^2 k^2\right]\psi = 0 \tag{3.11}$$

Differentiating our expression for ψ, we get

$$\frac{d^2\psi}{dx^2} = -A_0 4\pi^2 k^2 e^{2\pi i \mathbf{k} x} - 4\pi^2 \sum_{n\neq 0} A_n\left(k - \frac{n}{a}\right)^2 e^{2\pi i(\mathbf{k}-n/a)x} \tag{3.12}$$

Substituting ψ and $d^2\psi/dx^2$ in Eq. 3.11 and dividing through by $4\pi^2$, we get

$$\begin{aligned}\sum_{n\neq 0}\left[k^2 - \left(k - \frac{n}{a}\right)^2\right] A_n \exp\left\{2\pi i\left(\mathbf{k} - \frac{n}{a}\right)x\right\} \\ + \frac{2m}{h^2}\left(E - E_0 - \alpha\cos\frac{2\pi x}{a}\right) A_0 e^{2\pi i \mathbf{k} x} \\ + \frac{2m}{h^2}\left(E - E_0 - \alpha\cos\frac{2\pi x}{a}\right)\sum_{n\neq 0} A_n \exp\left\{2\pi i\left(\mathbf{k} - \frac{n}{a}\right)x\right\} = 0\end{aligned} \tag{3.13}$$

By using the orthogonality conditions we should now arrive at a set of relations between the A_n, but first we will make the job easier by some approximations.

If we assume that the interaction with the lattice is just a small perturbation on the free electron wave function, then α is small and likewise is $\Delta E = E - E_0$. This assumption also requires $\sum_{n\neq 0} A_n e^{2\pi i(\mathbf{k}-n/a)x}$ to be small, as may be seen from the assumed form of ψ.

With this assumption, we may neglect the last term, a product of two small terms, as compared with the other two

$$\begin{aligned}\sum_{n\neq 0}\left[k^2 - \left(k - \frac{n}{a}\right)^2\right] A_n e^{2\pi i(\mathbf{k}-n/a)x} \\ + \frac{2m}{h^2}\left(E - E_0 - \alpha\cos\frac{2\pi x}{a}\right) A_0 e^{2\pi i \mathbf{k} x} = 0\end{aligned} \tag{3.14}$$

To calculate the A_n, multiply by $e^{-2\pi i(\mathbf{k}-l/a)x}$ and integrate from 0 to a, where l is an integer $\neq 0$:

$$\begin{aligned}\sum_{n\neq 0}\left[k^2 - \left(k - \frac{n}{a}\right)^2\right] A_n \exp 2\pi i\left(\frac{l}{a} - \frac{n}{a}\right)x \\ + \frac{2m}{h^2}\left(E - E_0 - \alpha\cos\frac{2\pi}{a}x\right) A_0 e^{2\pi i(l/a)x}\end{aligned} \tag{3.15}$$

We now need some values of a few integrals:

$$\int_0^a e^{2\pi i(l/a)x}\,dx = 0 \tag{3.16}$$

$$\begin{aligned}\int_0^a \cos\frac{2\pi x}{a}\,e^{2\pi i(l/a)x}\,dx &= 0 \qquad \text{for } l \neq \pm 1 \\ &= \tfrac{1}{2}a \qquad \text{for } l = \pm 1\end{aligned} \tag{3.17}$$

$$\begin{aligned}\int_0^a \exp 2\pi i\left(\frac{l}{a} - \frac{n}{a}\right)x\,dx &= 0 \qquad \text{for } l \neq n \\ &= a \qquad \text{for } l = n\end{aligned} \tag{3.18}$$

For $l \neq \pm 1$:

$$a\left[k^2 - \left(k - \frac{l}{a}\right)^2\right]A_l = 0 = 2l\left(k - \frac{l}{2a}\right)A_l = 0 \tag{3.19}$$

Thus $A_n = 0$ for $n \neq \pm 1$, at least for $k \neq \pm l/2a$. For $l = +1$:

$$\left[k - \frac{1}{2a}\right]A_1 - \frac{2m\alpha}{h^2}\left(\frac{a}{2}\right)A_0 = 0 \tag{3.20}$$

or

$$A_1 = \frac{am\alpha}{2h^2(k - \frac{1}{2}a)}A_0 \tag{3.21}$$

Similarly,

$$A_{-1} = \frac{-am\alpha}{2h^2(k + \frac{1}{2}a)}A_0 \tag{3.22}$$

Thus

$$\psi = A_0\left\{e^{2\pi i kx} + \frac{\alpha m a}{2h^2}\left[\frac{e^{2\pi i(k-1/a)x}}{k - 1/2a} - \frac{e^{2\pi i(k+1/a)x}}{k + 1/2a}\right]\right\} \tag{3.23}$$

As long as $|k| \ll |1/2a|$, the last term is small and ψ does approximate the free wave function plus a small correction term.

Let us calculate $\Delta E = E - E_0$. Multiply Eq. 3.13 by $e^{-2\pi i kx}$ and integrate from 0 to a.

The first term contributes nothing.

The second term contributes $(2ma/h^2)(E - E_0)A_0$.

The third term gives

$$\frac{-2m\alpha}{h^2}\int_0^a (A_1 + A_{-1})\cos^2\frac{2\pi}{a}x\,dx = \frac{-am\alpha}{h^2}(A_1 + A_{-1})$$

Thus

$$\Delta E = (E - E_0) = \left(\frac{A_1 + A_{-1}}{2}\right)\alpha = \frac{m\alpha^2}{4h^2[k^2 - (1/4a)^2]} \tag{3.24}$$

As long as $|k| \ll |1/2a|$ this last term is small and E does not approximate E_0. Thus the assumptions we made in looking for a solution are justified by the solution as long as $|k| \ll |1/2a|$.

Now consider the case where $k \longrightarrow 1/2a$. Clearly A_1, A_0 become the important constants while the other A_n are small.

Thus

$$\psi = A_0 e^{2\pi i k x} + A_1 e^{2\pi i(k-1/a)x} \tag{3.25}$$

$$\frac{d^2\psi}{dx^2} = -4\pi^2 k A_0 e^{2\pi i k x} - 4\pi^2\left(k - \frac{1}{a}\right)^2 A_1 e^{2\pi i(k-1/a)x} \tag{3.26}$$

Introducing these into the Mathieu Eq. 3.2 and dropping $4\pi^2$, we get

$$A_0\left[\frac{2m}{h^2}\left(E - \alpha\cos\frac{2\pi x}{a}\right) - k^2\right]e^{2\pi i k x} + \left[\frac{2m}{h^2}\left(E - \alpha\cos\frac{2\pi x}{a}\right) - \left(k - \frac{1}{a}\right)^2\right]A_1 e^{2\pi i(k-1/a)x} = 0 \tag{3.27}$$

Multiply by $e^{-2\pi i k x}$ and integrate from 0 to a:

$$A_0\left[\frac{2m}{h^2}E - k^2\right]a - \frac{m\alpha}{h^2}aA_1 = 0 \tag{3.28}$$

Multiply by $e^{-2\pi i(k-1/a)x}$ and integrate from 0 to a:

$$-\frac{m\alpha}{h^2}aA_0 + a\left[\frac{2m}{h^2}E - \left(k - \frac{1}{a}\right)^2\right]A_1 = 0 \tag{3.29}$$

If a non-trivial solution for A_0, A_1 exists, then the determinate of the coefficients must vanish.

$$\left[\frac{2m}{h^2}E - k^2\right]\left[\frac{2m}{h^2}E - \left(k - \frac{1}{a}\right)^2\right] - \frac{m^2\alpha^2}{h^4} = 0 \tag{3.30}$$

When $k = 1/2a$ we get

$$\left(\frac{2m}{h^2}E - \frac{1}{4a^2}\right)\left(\frac{2m}{h^2}E - \frac{1}{4a^2}\right) - \frac{m^2\alpha^2}{h^4} = 0 \tag{3.31}$$

Thus
$$E = \frac{h^2}{8ma} \pm \frac{\alpha}{2} = E_0 \pm \frac{\alpha}{2} \tag{3.32}$$

Using this value for E, we get $A_1 = \pm A_0$ at $k = 1/2a$ or

$$\psi = A(e^{i\pi x/a} \pm e^{-i\pi(x/a)}) \tag{3.33}$$

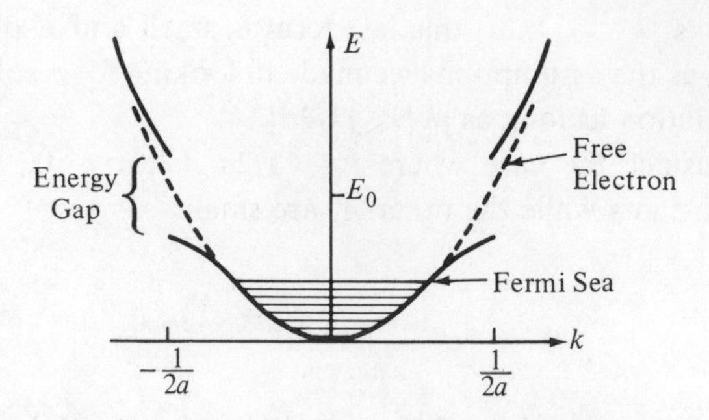

Figure 3.1 Plot of Energy of Electron as a function of Wave Vector **k**.

Thus we see that at $k = 1/2a$ we have an energy gap, see Fig. 3.1.

Another important approach to the one-dimensional electron wave in a metal was developed by Hubert M. James of Purdue University.[2] We again consider the Schrödinger wave equation for an electron moving in a symmetric potential as shown in Fig. 3.2*a*, considering the energy as constant at some particular value but not in the forbidden energy gap region. The wave equation has two linearly independent solutions within each cell described in Fig. 3.2*a*. Consider the zeroth cell between $0 \leq x \leq a$ and the two independent solutions may be represented by $g(x)$ and by $u(x)$ as shown in Figs. 3.2*b* and 3.2*c*.

The properties of these even and odd wave functions are as follows:

$$\begin{aligned}
g(0) &= g(a) \quad &\text{and} \quad &\frac{dg(0)}{dx} = -\frac{dg(a)}{dx} \equiv g'(a) \\
g(a/2) &= 1 \quad &\text{and} \quad &\frac{dg(a/2)}{dx} = 0 = g'(a/2) \\
u(a/2) &= 0 \quad &\text{and} \quad &\frac{du(a/2)}{dx} = 1 = u'(a/2) \\
u(0) &= -u(a) \quad &\text{and} \quad &\frac{du(0)}{dx} = \frac{du(a)}{dx} = u'(a)
\end{aligned} \tag{3.34}$$

The complete wave function in the zeroth cell is the sum of the particular solutions.

$$\psi_0(x) = \alpha_0 g(x) + \beta_0 u(x) \tag{3.35}$$

where the amplitudes α_0 and β_0 have been used.

[2]H. M. James, *Phys. Rev.* **76**, 1602, 1949.

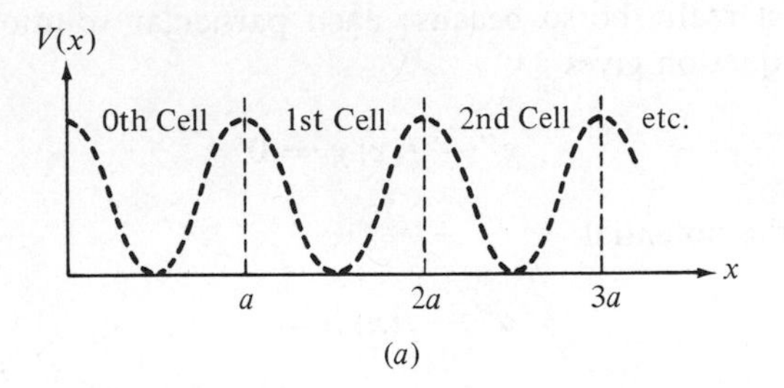

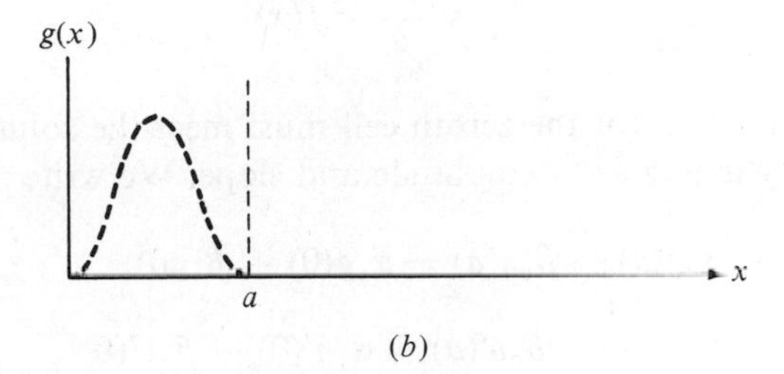

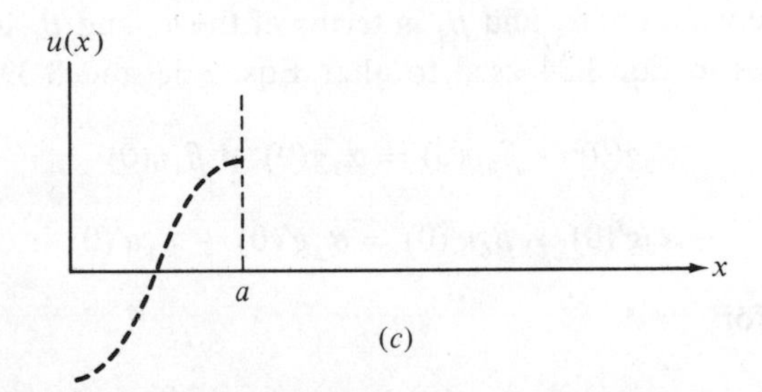

Figure 3.2a Plot of Potential on Electron in each cell as a function of x. **b** Even function solution to wave equation in cell. **c** Uneven function solution to wave equation in cell.

The Wronskian is constant in the interval and, in fact, at the midpoint $a/2$ the Wronskian equals unity. To see that this is so we write the Wronskian

$$\begin{vmatrix} g(x) & u(x) \\ g'(x) & u'(x) \end{vmatrix} = \text{constant} \tag{3.36}$$

Expanding and taking the derivative we have

$$[g(x)u'(x) - g'(x)u(x)]' = g'(x)u'(x) + g(x)u''(x) - g'(x)u'(x) - g''(x)u(x)$$

and this is set equal to zero because the derivative of a constant is zero. Thus the equation becomes

$$g(x)u''(x) - g''(x)u(x) = 0 \qquad \text{or} \qquad \frac{g''}{g} = \frac{u''}{u} \tag{3.37}$$

Now this must really be so because each particular solution put into the Schrödinger equation gives

$$g'' - f(x)\, g = 0$$

where $f(x)$ is the potential

and
$$u'' - f(x)\, u = 0$$

so that
$$\frac{g''}{g} = \frac{u''}{u} = f(x) \tag{3.37'}$$

The solution at $x = a$ for the zeroth cell must meet the solution in the first cell so that they match as to amplitude and slope. We write

$$\alpha_0 g(a) + \beta_0 u(a) = \alpha_1 g(0) + \beta_1 u(0) \tag{3.38}$$

$$\alpha_0 g'(a) + \beta_0 u'(a) = \alpha_1 g'(0) + \beta_1 u'(0) \tag{3.39}$$

We can get the values of α_1 and β_1 in terms of the α_0 and β_0 in the following way (see values in Eq. 3.34 used to alter Eqs. 3.38 and 3.39):

$$\alpha_0 g(0) - \beta_0 u(0) = \alpha_1 g(0) + \beta_1 u(0) \tag{3.38'}$$

$$-\alpha_0 g'(0) + \beta_0 u'(0) = \alpha_1 g'(0) + \beta_1 u'(0) \tag{3.39'}$$

Solving 3.38′ for

$$\beta_1 = \alpha_0 \frac{g(0)}{u(0)} - \beta_0 - \alpha_1 \frac{g(0)}{u(0)}$$

and substituting into 3.39 gives

$$\alpha_0[g'(0)u(0) + g(0)u'(0)] - 2\beta_0 u'(0)u(0) = \alpha_1 \tag{3.40}$$

where we have used the fact that the Wronskian

$$[g(0)u'(0) - g'(0)u(0)] = 1$$

Thus we see the scheme for calculating the amplitudes for the first cell becomes in matrix algebra notation

$$\begin{pmatrix} \alpha_1 \\ \beta_1 \end{pmatrix} = \begin{pmatrix} g'(0)u(0) + g(0)u'(0) & -2u'(0)u(0) \\ -2g(0)g'(0) & g'(0)u(0) + g(0)u'(0) \end{pmatrix} \begin{pmatrix} \alpha_0 \\ \beta_0 \end{pmatrix}$$

Or in a shorthand notation of symbols this becomes

$$\begin{pmatrix}\alpha_1\\ \beta_1\end{pmatrix} = \begin{pmatrix}A_{11} & A_{12}\\ A_{21} & A_{22}\end{pmatrix}\begin{pmatrix}\alpha_0\\ \beta_0\end{pmatrix} \tag{3.41}$$

This relationship is often used as a starting point in discussing the amplitude change of a wave moving in a periodic structure. If the wave amplitude does not change by translation of a distance *na* from the zeroth cell to the *n*th cell, then $\psi(x + na) = e^{2\pi i \mathbf{k}\cdot na}\psi(x)$ and we have changed the wave only by a simple phase factor.[3]

We shall continue the James[2] discussion in order to outline a rigorous proof of this last statement. Let the energy, E, be taken as a constant and let the wave function be $f(x) = \alpha g(x) + \beta u(x)$ in the zeroth cell. Then let the wave function be $rf(x - a)$ in the first cell, $r^2 f(x - 2a)$ in the second cell; $\cdots r^n f(x - na)$ for $na < x < (n + 1)a$. The quantity $r = f(a)/f(0)$ which is the ratio of the amplitude of the wave function at the end of the zeroth cell to that at its beginning. For a boundary fit which demands the same slope-to-magnitude ratio at the two ends of the cell we write

$$\frac{\alpha g'(0) + \beta u'(0)}{\alpha g(0) + \beta u(0)} = \frac{\alpha g'(a) + \beta u'(a)}{\alpha g(a) + \beta u(a)} = \frac{-\alpha g'(0) + \beta u'(0)}{\alpha g(0) - \beta u(0)} \tag{3.42}$$

where we have again used the values of Eq. 3.34. From this last equation we obtain

$$\left(\frac{\alpha}{\beta}\right)^2 = \frac{u(0)u'(0)}{g(0)g'(0)} = \left(\frac{u(0)}{g(0)}\right)^2 \cdot \frac{1}{\rho} \tag{3.43}$$

where we have defined the symbol

$$\rho \equiv \frac{g'(0)u(0)}{g(0)u'(0)} \tag{3.44}$$

In these symbols the multiplying quantity, r, becomes

$$r = \frac{\alpha g(a) + \beta u(a)}{\alpha g(0) + \beta u(0)} = \frac{\alpha g(0) - \beta u(0)}{\alpha g(0) + \beta u(0)} = \frac{1 - (\beta/\alpha)[u(0)/g(0)]}{1 + (\beta/\alpha)[u(0)/g(0)]} \tag{3.45}$$

In terms of the quantity, ρ, the multiplying factor is

$$r_\pm = \frac{1 \mp \rho^{1/2}}{1 \pm \rho^{1/2}} \tag{3.46}$$

[3] A very appealing presentation of this property has been given by W. A. Harrison, *Physics Today*, **22**, 10, p. 23 (1969) on the occasion of a symposium in honor of the 40th anniversary of F. Bloch's thesis publication. Harrison calls the potential in Eq. 3.2 a "pseudopotential" and the resulting wave function, Eq. 3.9, a "pseudowavefunction."

with the signs $\pm$ depending upon taking the square root in Eq. 3.43. The general solution in the nth cell is a linear combination with real coefficients C_+ and C_- which gives

$$\psi(x) = C_+ r_+^n f_+(x - na) - C_- r_-^n f_-(x - na) \tag{3.47}$$

We demand that this function be finite and reasonably "well behaved," analytically speaking. We examine what physically acceptable solutions we get for the following possibilites: $\rho > 0$ or $\rho < 0$ as well as the trivial $\rho = 0$ and $\rho = \infty$. It turns out[2] that only for $\rho < 0$ do we get a finite, non-trivial solution. For negative values of ρ this means: $\rho^{1/2}$ is imaginary, β/α is imaginary, $f_\pm(x)$ is complex, and

$$r_\pm = \frac{1 \mp i\sqrt{\rho}}{1 \pm i\sqrt{\rho}} \equiv e^{\pm i\theta} \tag{3.48}$$

Thus the functions $r_+^n f_+(x - na)$ and $r_-^n f_-(x - na)$ change phase from cell to cell, but do not increase in magnitude. If we translate by a lattice vector $\mathbf{l} = na$, the wave function is multiplied by a phase factor $e^{i\mathbf{k}_1 \cdot \mathbf{l}}$ as follows:

$$\psi_{\mathbf{k}_1}(\mathbf{r} + \mathbf{l}) = e^{i\mathbf{k}_1 \cdot \mathbf{l}} \psi_{\mathbf{k}_1}(\mathbf{r}) \tag{3.49}$$

where $\mathbf{k}_1 = 2\pi\mathbf{k}$ in the above expression will remind us that the wave vector used in this chapter, e.g., Eq. 3.3, is sometimes written without the factor 2π and in this case the energy of the free electron is $E = k_1^2\hbar^2/2m$.

3.1 PHYSICAL INTERPRETATION OF ELECTRON WAVES IN SOLIDS

The main theme of this book is to discuss waves in a number of simple physical systems. The theory becomes sterile if we do not link up with physical reality and there is importance in useful information calculated or made clear with the theory. Thus we briefly digress into the physics of metals, semiconductors, and insulators[4] to which the above theory relates. We have seen that the edge of the first Brillouin zone occurs when the wave vector $\mathbf{k} = 1/2a$ where a is the distance between the atoms. For copper the nearest neighbor is 2.55×10^{-8} cm so, in that direction, the value of $k_{max} = 1.96 \times 10^7$ cm^{-1} at the top of the first allowed energy band shown in

[4] A first-year graduate course in solid-state physics, based on the sort of wave mechanics given in this chapter, might be, for example: *Principles of The Theory of Solids*, by J. M. Ziman, Cambridge University Press, 1964.

Fig. 3.1. If the free electron energy, $E_0 = k^2h^2/2m$, were calculated from this maximum value of the wave vector, we obtain 9.27×10^{-12} erg for an electron. This is equivalent to 5.8 electron-volts or to a maximum speed of 1.4×10^8 cm/sec. For an electron treated as a gas in classical kinetic theory, the temperature required to give an average velocity of 1.4×10^8 cm/sec would be a staggering 65,000 Kelvin! Even the bending down of the actual curve in Fig. 3.1 does not greatly lower the energy (and hence the velocity) of these electrons. This is true even at 0 Kelvin. Incidentally, copper is a face-centered cubic solid structure of cubic edge 3.61×10^{-8} cm. The value of k_{max} computed for a k-space direction corresponding to a cubic edge would then be slightly different. Roughly, the three-dimensional figure of k_{max} is the surface of a sphere.

Introductory courses about the atom usually portray electrons orbiting a small core of positive charge such that electrostatic attraction is balanced by the centrifugal force and the linear velocity of a circular orbiting electron is of the magnitude 5×10^8 cm/sec (see Chapter 8). We consider the single, gas-like, copper atom as a tight central core made up of the nucleus and of a large number of closely held electrons. A single outer orbiting or valence electron moves with a speed of some 2×10^8 cm/sec. When the metal is formed by the atoms condensing into the solid state, the outer electron of each atom may retain much of its former velocity and may move through the lattice in the voids between the atoms. Each copper atom contributes one "free" electron and these belong to the crystal as a whole. Each electron is described by the wave Eq. 3.2. Of course these N electrons coming from the N atoms of copper are dashing about like wild animals in a cage unable to escape at the surface of the metal because of the binding between the electron and the ionic core of the surface atoms. A little help from an impinging photon or a tug from an external electric field can assist in the escape of the most rapid electrons to give electron emission.

The number of electrons in the first band of energy shown in Fig. 3.1 is limited because the standing waves, as in the case of the string, have definite wave lengths: $\lambda = 2l/n = 2Na/n$. The wave vectors are thus limited to special values: $|k| = 1/\lambda = n/2aN$ in the direction of the interatomic distance a and where n is an integer or quantum number. The first allowed band has values of k from 0 up to $1/2a$ according to the theory of this chapter. Thus the integers n can run up to N and there are but N energy states available to the N free electrons.

Why should it not be possible for all N electrons to lose energy and occupy the lowest possible energy state corresponding to a quantum number $n = 1$ or a wave vector $\mathbf{k} = 1/2aN$? Of course we could have asked such a question of the valence electron orbiting the single atom and we know that exclusion laws prohibit the electron from falling into lower orbits. It is shown in a more advanced course in quantum wave mechanics that electrons of opposite spin angular momentum (a new concept?) may occupy an energy state two at a time but no more than two. We think of the energy states like an apartment building having one apartment on each floor and in each apartment two opposite spin electrons may dwell. At the top of the building the apartments are empty because most of the N electrons are in pairs in the bottom $N/2$ apartments. A few batchelor electrons, or single electrons, may dwell alone in the apartments just above (and having been deserted by their partners also just below) the $N/2$ floor level. Thus for copper the available N electrons at room temperature never fill up all N energy states up to the first Brillouin zone where $\mathbf{k} = 1/2a$. Even so, the electrons at the level corresponding to $n = N/2$ or to $\mathbf{k} = 1/4a$ are fast-moving electrons of the order 10^8 cm/sec. These are the relatively few electrons which can be excited by thermal agitation to move to an empty energy state corresponding to, say, $n = 1 + N/2$ and which is ever so little higher in energy. Those happy couples of electrons living in the lower levels of the apartiment building do not participate in the thermal energy of excitation because they do not have a vacant apartment close by. For these reasons the specific heat of the electron gas at room temperature is negligible compared to the lattice vibration specific heat. The energy of N electrons is $U = U_0 + 1/2\alpha T^2$ where T is the absolute temperature. The specific heat is $C_e = \partial U/\partial T$ and we have given a picture above that U_0 is large but α is small. Electrons obey a quantum statistics (Fermi-Dirac) which is the real scientific label behind the above analogy to apartment house living rules for electrons. The electron theory of metals was developed by Sommerfeld and his students F. Bloch, W. V. Houston, and others.

We have deliberately chosen copper, a monovalent metal, so that the first band would be only about half full and empty energy states would be readily available to an electron at the top of the k-space sphere, or top of the "Fermi Sea" as the jargon says. Application of a voltage across the metal means that the electrons at the surface of the sphere react to the electric field, acquire more speed, and move to an upper, slightly higher, and empty energy state. Thus a current flows in the direction of increased k and this is characteristic of a metal. But if we had chosen a divalent atom to form the solid state, the first allowed energy band could be filled with the $2N$ electrons occupying the N available energy states. There could then be an energy gap above the filled band as shown in Fig. 3.1 and as given in the above theory. A small electric field would not be able to excite the upper electrons of the filled band to cross the energy gap so the solid would act as an insulator. Not all divalent atoms are necessarily an insulator because the upper band in Fig. 3.1 might overlap the bottom band and produce a metal—at least in certain directions in **k**-space. A semiconductor is characterized by a small energy gap so that a few electrons are thermally excited across the gap to the upper band where they can, in turn, produce a net electric current if an electric field is applied. All of these explanations of the properties of solids have been based on the wave properties of electrons moving through the periodic potential of the lattice as introduced in this chapter. Obviously, there are many more interesting properties which these so called pseudowave-functions and pseudopotentials can be made to explain; superconductivity for one.[5]

[5]P. B. Allen and M. L. Cohen, *Phys. Rev.*, **187**, 525 (1969).

Chapter 4

WAVE PROPAGATION IN ONE DIMENSION

4.1 TRANSMISSION LINES

In this chapter we seek to capture the general idea of wave propagation. Waves behave in a similar way and by giving the mechanical properties of a transmission line which is again the stretched string, the methods will be set up for considering other elastic, electromagnetic, and optical waves in later chapters. At one end of the string we place the energy source or oscillator as shown in Fig. 4.1. The string is stretched with a tension, T, and has mass per

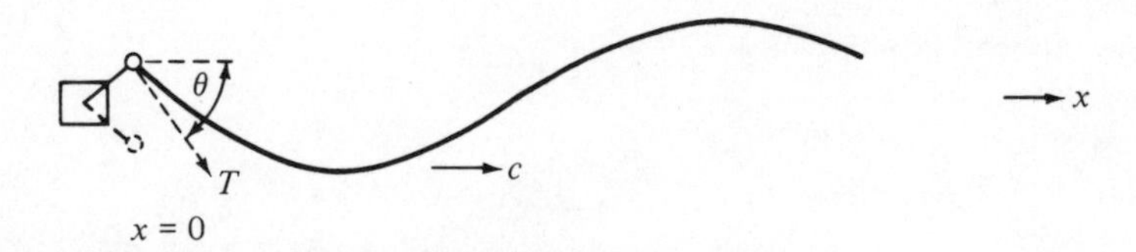

Figure 4.1 Flexible String as Infinite Transmission Line.

unit length, ρ. Let us take the string to be very long so that when we first start up the oscillator we see waves moving out with a velocity, c, and no return waves. A theoretical scientist would say let us take an infinitely long string in order to have no return waves, but we do not demand such imagination. Perhaps the string could really have the transverse wave amplitude get slightly smaller due to friction as it moves out, and after a very long way to the other end there is very little energy in the wave to reflect. Or even better, the support at the far end could be so cleverly designed as to just take up the incident wave energy in its own damped oscillating system such that there is no reflected wave on the string; i.e., the string acts like an infinite transmission line. At $x = 0$ the force on the string by the oscillator is $F_y = F_0 e^{j\omega t}$ and the y-component of force by the string on the oscillator arm is opposite; i.e.,

$$-T \sin \theta \simeq -T \tan \theta = -T\left(\frac{\partial y}{\partial x}\right)_{x=0}$$

Thus we set

$$F_0 e^{j\omega t} = -T\left(\frac{\partial y}{\partial x}\right)_{x=0} \tag{4.1}$$

The wave moving out to the right on the string is one which must obey the wave Eq. 2.3 of Chapter 2 and by Eq. 2.8 the solution

$$y = C_+ e^{-j(\omega x/c)} \cdot e^{j\omega t}$$

is the required displacement. There is no friction or damping of the wave. Differentiation with respect to x and evaluated at $x = 0$ we have

$$\left(\frac{\partial y}{\partial x}\right)_{x=0} = -\frac{j\omega}{c} C_+ e^{j\omega t} \tag{4.2}$$

Then the upward force of F_0 equals the downward force of the string to give the amplitude relation

$$F_0 = T \frac{j\omega C_+}{c} \tag{4.3}$$

The outgoing wave is written with the newly determined amplitude

$$y = \frac{cF_0}{j\omega T} e^{-j(\omega x/c)} \cdot e^{j\omega t} \tag{4.4}$$

Differentiation with respect to time gives the velocity

$$v = \frac{\partial y}{\partial t} = \frac{cF_0}{T} e^{-j(\omega x/c)} \cdot e^{j\omega t} \tag{4.5}$$

At $x = 0$ the ratio of the force to the velocity becomes

$$\frac{F_0 e^{j\omega t}}{v} = \frac{T}{c} \equiv Z_0 \tag{4.6}$$

which is the characteristic mechanical impedance of the lossless infinite string. Since $c = \sqrt{T/\rho}$, we may also write the input impedance $Z_0 = \rho c$ which is a real number and not complex. Every lossless medium which propagates a wave has a characteristic impedance of this type.

If the string is of finite length, l, and it is driven at $x = 0$ as before, but is supported at $x = l$ with a mechanical device, the input impedance is quite different because of the reflected waves. Figure 4.2*a* shows the outgoing wave and the reflected wave. These two waves add at all times to give a motion contained within the wave envelope shown in Fig. 4.2*b*. The actual string

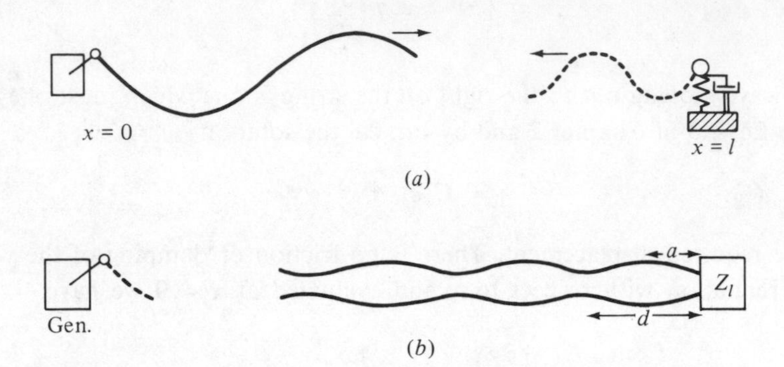

Figure 4.2a Flexible String as Transmission Line with Terminating Impedance. **b** Standing Wave Pattern on Transmission Line.

at any point x moves up and down across $y = 0$ with a frequency ω, but the amplitude of the so-called standing wave pattern is different at various points x. The amplitude is not necessarily at all times equal to zero at what looks like nodal points, for these are only minimum amplitude points.

Using the solution to the wave equation obtained in Chapter 2, Eq. 2.10, we have an expression which gives the space and time variation

$$y = C_+ e^{j\omega(t-x/c)} + C_- e^{j\omega(t+x/c)} \tag{2.10'}$$

Unlike Eq. 2.11 relating the amplitude and phase of the outgoing and incoming wave on the string, the reflected wave is down in amplitude and shifted by a different phase angle so that we write at the origin $x = 0$, the new expression

$$C_- = C_+ e^{-2\pi\alpha_0} \cdot e^{-j2\pi\beta_0} = C_+ e^{-2\pi\alpha_0 - j2\pi\beta_0} \tag{4.7}$$

The transmission line itself has no damping term to produce wave amplitude attenuation. The exponential with $(-2\pi\alpha_0)$ lowers the amplitude of the reflected wave with respect to the outgoing wave, and this is taken to be caused by the damping action at the end support or load impedance. The exponential with $(-j2\pi\beta_0)$ is a simple phase angle shifter and we shall now see how these quantities are determined. Substitution into the space-time equation for the displacement gives

$$y = C_+ e^{j(\omega t - \mathbf{k}x)} + C_+ e^{-2\pi\alpha_0 - j2\pi\beta_0} e^{j(\omega t + \mathbf{k}x)} \tag{4.8}$$

or

$$y = 2C_+ e^{-\pi\alpha_0 - j\pi\beta_0}\left[\frac{e^{-j\mathbf{k}x + \pi\alpha_0 + j\pi\beta_0} + e^{-\pi\alpha_0 - j\pi\beta_0 + j\mathbf{k}x}}{2}\right] e^{j\omega t} \tag{4.9}$$

$$y = C \cosh\left[\pi(\alpha_0 + j(\beta_0 - kx))\right] e^{j\omega t} \tag{4.10}$$

This last equation gives the space-time variation of the displacement of the string with functions of space variation slightly more complicated than those used in Eq. 2.14 of Chapter 2, and with complex amplitude and time variation somewhat similar but not identical to those used in Eq. 2.14 of Chapter 2.

Out at the load end of the driven string, at $x = l$, we may now use our theory to calculate the force on the string, the displacement velocity, and finally the complex impedance Z_l at that point. The force on the string at $x = l$ becomes

$$F_l = -T\left(\frac{\partial y}{\partial x}\right)_{x=l} = -TC\{-jk\} \sinh\left[\pi\left(\alpha_0 + j\left(\beta_0 - \frac{kl}{\pi}\right)\right)\right] e^{j\omega t} \tag{4.11}$$

The velocity is

$$v_l = \left(\frac{\partial y}{\partial t}\right)_{x=l} = j\omega C \cosh\left[\pi\left(\alpha_0 + j\left(\beta_0 - \frac{kl}{\pi}\right)\right)\right] e^{j\omega t} \tag{4.12}$$

and the impedance is the ratio of force to velocity at the load

$$Z_l = \rho c \tanh \pi\left[\alpha_0 + j\left(\beta_0 - \frac{kl}{\pi}\right)\right] \tag{4.13}$$

Using the identity $k = 2\pi/\lambda$ where λ is the wavelength on the string, we may also write

$$Z_l = \rho c \tanh \pi\left[\alpha_0 + j\left(\beta_0 - \frac{2l}{\lambda}\right)\right] \tag{4.14}$$

The general expression for the impedance at any point x along the string is

$$Z(x) = \rho c \tanh \pi\left[\alpha_0 + j\left(\beta_0 - \frac{2x}{\lambda}\right)\right] \tag{4.15}$$

The pure imaginary term $j(\beta_0 - 2x/\lambda)$ can be written as simply $j\beta(x)$ and specifically the value $\beta_l = \beta_0 - 2l/\lambda$ may be used in the expression for the complex impedance at $x = l$

$$Z_l = R_l + jX_l = \rho c \tanh \pi(\alpha_0 + j\beta_l) \tag{4.16}$$

The value of α_0 is the same at all points x including $x = l$. The impedance at the load end is made up of a resistive part and a reactance part, just as in electrical transmission lines. In fact, the string as a transmission line is so similar to the electric transmission line that we have used the notation of the electrical engineer with $j = -i$ in the pure imaginary quantities. Returning to Eq. 4.10, for the displacement we use some functional identities to obtain

$$y = C \cosh\left[\pi\left(\alpha_0 + j\left(\beta_0 - \frac{2x}{\lambda}\right)\right)\right] e^{j\omega t} \tag{4.10'}$$

or

$$y = C\left\{\cosh \pi\alpha_0 \cos \pi\left(\beta_0 - \frac{2x}{\lambda}\right) + j \sinh \pi\alpha_0 \sin \pi\left(\beta_0 - \frac{2x}{\lambda}\right)\right\} e^{j\omega t} \tag{4.17}$$

and

$$|y| = |C| \sqrt{\cosh^2 \pi\alpha_0 - \sin^2 \pi(\beta_0 - 2x/\lambda)} \tag{4.18}$$

gives the displacement magnitude.

We get *minimum* when

$$\sin^2 \pi\left(\beta_0 - \frac{2x}{\lambda}\right) = 1 \tag{4.19}$$

We get *maximum* when

$$\sin^2 \pi\left(\beta_0 - \frac{2x}{\lambda}\right) = 0 \tag{4.20}$$

Ratio

$$\frac{|y|_{\min}}{|y|_{\max}} = \frac{\sqrt{\cosh^2 \pi\alpha_0 - 1}}{\sqrt{\cosh^2 \pi\alpha_0}} = \tanh \pi\alpha_0 \tag{4.21}$$

so

$$\alpha_0 = \frac{1}{\pi} \tanh^{-1} \frac{|y|_{\min}}{|y|_{\max}} \tag{4.22}$$

allows us to calculate α_0 from the standing wave ratio we have measured. Now minimum occurs when

$$\sin \pi\left(\beta_0 - \frac{2x}{\lambda}\right) = \pm 1 \tag{4.23}$$

so

$$\pi\left(\beta_0 - \frac{2x}{\lambda}\right) = \frac{(2n + 1)\pi}{2} \quad \text{or} \quad \beta_0 = n + \frac{1}{2} + \frac{2x}{\lambda} \tag{4.24}$$

Likewise, since $\beta_0 = \beta_l + 2l/\lambda$,

$$\beta_l + \frac{2l}{\lambda} = n + \frac{1}{2} + \frac{2x}{\lambda} \tag{4.25}$$

where x is the distance to a minimum.

Solving we obtain

$$\beta_l = n + \frac{1}{2} - \frac{2}{\lambda}(l - x) = n + \frac{1}{2} - \frac{d}{\lambda/2} \tag{4.26}$$

If d, the distance from $x = l$ to the closest minimum, is greater than $\lambda/4$, n must be taken as 1 to get a value of β_l between 0 and 1. If d is less than $\lambda/4$, n can be taken as 0. The distance d is shown in Fig. 4.2*b*.

Similar relations can be obtained for the position of the maxima with respect to the ends of the string. The points of maximum amplitude are one-half wavelength apart and the distance between neighboring minima and maxima is one-quarter wavelength.

The situation can be illustrated as shown in Fig. 4.3.

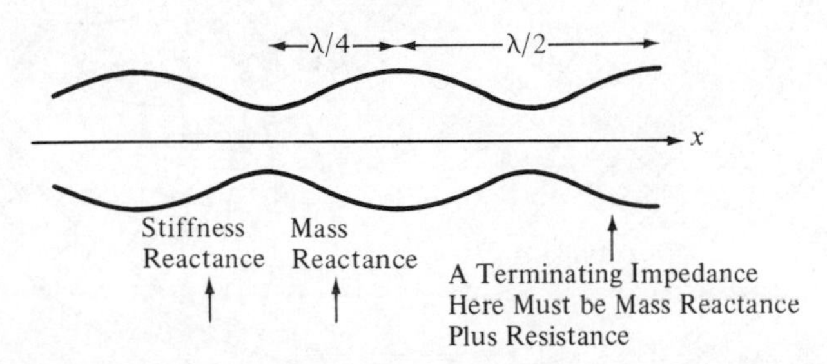

Figure 4.3 Standing Wave Pattern on Transmission Line showing Impedance, $Z(x)$.

4.2 SMITH CHART

The calculation with hyperbolic functions of complex arguments can be done with some labor. The introduction by P. H. Smith, of the Bell Telephone Laboratories, of a chart[1] shown in Fig. 4.4 has made the calculation less tedious. Individual sheets of the Smith Chart can be purchased for a few pennies at any university book store because most Electrical Engineering Departments require these in working problems on microwave transmission lines, and these same sheets can be used for problem solving in this text. The procedure is as follows:

[1] *Electronics*, **17**, 130-133 and 318-325 (1944). See also, *Electronic Applications of the Smith Chart*, by Phillip H. Smith, McGraw-Hill Book Co. (1969).

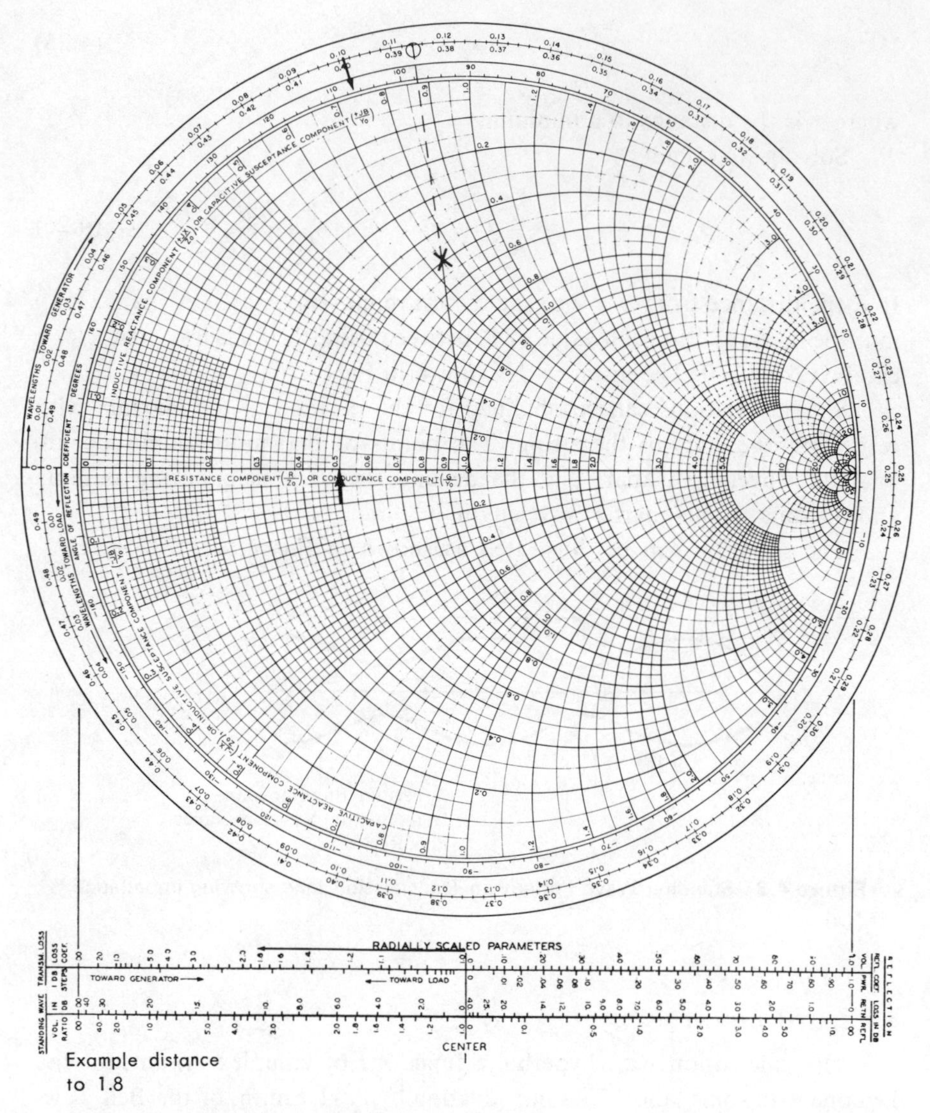

Figure 4.4 Smith Chart showing Impedance Coordinates, reproduced by permission of Kay Electric Company, Pine Brook, New Jersey.

1. Measure the standing wave ratio on the string which is $|y|_{max}$ divided by $|y|_{min}$ and this number (e.g., take 1.80) determines the distance on the Radially Scaled Parameters seen at the bottom and marked "Standing Wave VOL. Ratio." Mark that distance off on the straight edge of a sheet of $8\frac{1}{2} \times 11$ inch paper. See distance to 1.80 shown on Fig. 4.4 as example.

2. Measure the distance a from the load end at $x = l$ back to the first *maximum* on the string. Express this distance as the number of wave lengths, a/λ, that the first maximum occurs from the load towards the generator. The number is a fraction between 0 and $\frac{1}{2}$ and this fraction is now located on the outer circumference of the Smith Chart marked "Wavelengths Toward Generator—." In example suppose this is 0.114 marked by circle on Fig. 4.4.
3. Put on top of the Smith Chart the sheet of paper (with the standing wave ratio distance parameter scaled on it) such that the edge of the paper touches the proper circumference fraction determined in step 2 and the edge also touches the very center of the Smith Chart at the point marked 1.0 on the axis marked "Resistance Component." Slide the edge of the paper along such that the distance parameter determined in step 1 measures radially out from the center towards the proper fraction number determined in step 2. The point on the Smith Chart which is the distance out on the standing wave parameter is now marked as the locus of the real and pure imaginary components, R/Z_0 and $\pm jX/Z_0$, of the impedance Z_l. See point marked x for example taken on Fig. 4.4.
4. Follow the circle (which passes through the point determined) back to the resistance component axis and read off the value R/Z_0. Interpolate between circles if needed. See arrow on Fig. 4.4 for example.
5. Follow the curves which fan out radially (and which contain the point) to read off the reactive component ratio $\pm jX/Z_0$. These values are located out at the circumference of the Smith Chart. Interpolate between curves if needed. See arrow on Fig. 4.4 for example.
6. The actual impedance of the load or support mechanism of the string at $x = l$ is thus determined from a knowledge of the characteristic line impedance Z_0 and from the values located on the Smith Chart.

The theoretical or mathematical background for the Smith Chart is not difficult, but would require some pages of discussion. The background for the Smith Chart may be found in many texts which specialize in the transmission line theory; e.g., *Microwave Transmission* by J. C. Slater, McGraw-Hill Book Co., New York (1942) and more recently *Theory and Problems of Transmission Lines* by R. A. Chipman, Schaum's Outline Series of McGraw-Hill Book Co., New York (1968). Of course we refer to the original article by Smith and his more recent book[1]. The circles represent values of constant α but the circles do not have a common center on the resistance component axis. The segments which fan out from a common point are constant values of β and these segments are parts of circles whose centers are located on an axis at right angles to the resistance component axis at the extreme right of the chart. Ordinary transmission lines have positive real-part impedances as we have discussed here but in electrical work it is possible to have negative real-part impedances; e.g., when an Esaki diode is used. The rules for using the Smith Chart must then be changed to suit the new situation of a negative real-part impedance.

PROBLEMS

1. A string driven from the left end at $x = 0$ by a sinusoidal force is terminated at the right end at $x = l$ with a mechanical device of complex impedance. It is found that the string has a mass of 0.1 gm per cm of length and is under a tension of 10^5 dynes. The distance from the load at $x = l$ back to the first maxima in the envelope of the transverse displacement is 6 cm. The maxima are observed to be 10 cm apart. The ratio of $|y|_{max}$ to $|y|_{min}$ is measured to be 1.80. What is the characteristic impedance of the string? What is the mechanical impedance of the load, Z_l?

2. The basic transmission line circuit for electromagnetic wave propagation is shown in the following diagram:

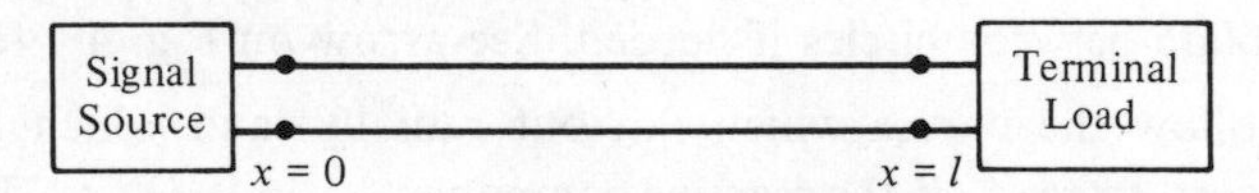

The transmission line is a uniform, lossless system and the terminal load is a complex electrical impedance, Z_l. The standing wave in voltage caused by the terminal load impedance might look as follows:

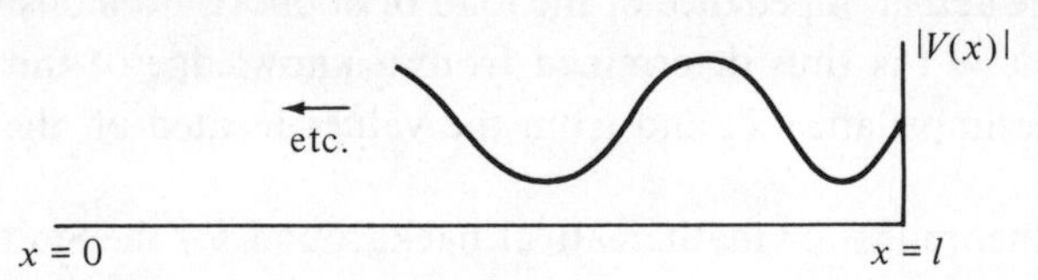

The lossless transmission line has a characteristic impedance of $50 + j0$ ohms. There is a voltage standing wave ratio (VSWR) of 4.0. The distance between successive minima of the pattern is 50 cm, and there is a voltage minimum 40 cm from the load terminals. Determine the terminal load impedance connected to the line. Use the Smith Chart for this problem.

Chapter **5**

WAVES IN TWO DIMENSIONS: The Flexible Membrane

This chapter serves to enrich our understanding of wave mechanics by considering the two-dimensional system called a flexible membrane. As in the flexible string, the stiffness is treated as negligible in order to idealize the physical situation. The mathematical tools are developed to describe the observations. In the one-dimensional cases treated in the earlier chapters some practical examples in both classical and quantum systems could be studied with the techniques developed. In the two-dimensional wave propagation examined in the present chapter, wc make no pretence that the membrane can be closely identified with real systems; e.g., drumheads, microphones, and telephone receiver plates. We must appeal to imagination as to how the flexible membrane would oscillate in its characteristic modes and how it would respond to a driving force. The aim is to develop new concepts and mathematical tools rather than to explain in depth the behavior of real membranes. The justification is that these will be useful as background in treatment of three-dimensional wave propagation, scattering, and other characteristics.

5.1 THE RECTANGULAR SHAPED FLEXIBLE MEMBRANE

Figure 5.1*a* is a diagram of the rectangular flexible membrane stretched between rigid supports with a tension, T, which is equal in magnitude on all sides and expressed in dynes per centimeter of length along the side. The

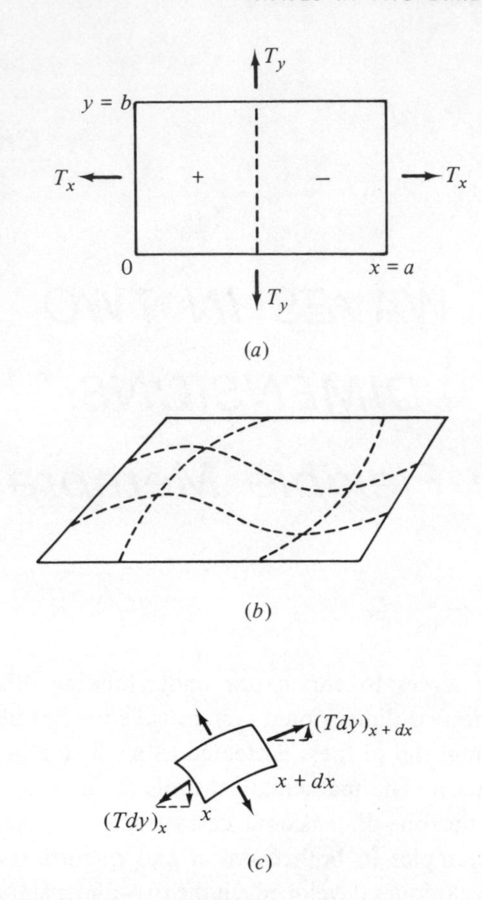

Figure 5.1a Rectangular Membrane. **b** Diagram of Mode of Motion. **c** Segment of Rectangular Membrane showing Forces.

displacement of this membrane can be very complicated if it is set in motion by striking an area in the center or on the right lower quadrant for example. We can imagine that had we pushed upward with our hand on the left portion of Fig. 5.1*a* and pushed downward on the right portion, we might set into motion a characteristic mode of vibration with a nodal line across the center as shown by the dotted line. Such a motion could be given a perspective view by an artist as shown in Fig. 5.1*b* with lines representing the motion of the membrane which would be like a two-dimensional set of flexible strings. The string in the x-direction has a nodal point and is in its second harmonic mode. The y-direction string is in its fundamental mode of motion. It will be no surprise that a theoretical analysis of the rectangular flexible membrane will appear as a simple two-dimensional treatment of the flexible string treated in Chapter 2.

The element of membrane $dx\,dy$ shown in Fig. 5.1c has a mass per unit area ρ grams per cm^2. The equation of motion must add up the displacement forces arising from tension in the x-direction with tension in the y-direction and set the resulting vertical force equal to the mass times the acceleration of the segment in the third dimension of displacement, ψ. Using the similar derivation for the flexible string of Chapter 2, we now have simply

$$\left(\frac{\partial^2\psi}{\partial x^2} + \frac{\partial^2\psi}{\partial y^2}\right) T\,dx\,dy = \rho\,dx\,dy \frac{\partial^2\psi}{\partial t^2} \tag{5.1}$$

or

$$\frac{\partial^2\psi}{\partial x^2} + \frac{\partial^2\psi}{\partial y^2} = \frac{\rho}{T}\frac{\partial^2\psi}{\partial t^2} \tag{5.2}$$

If we define the wave velocity $c = \sqrt{T/\rho}$, we may write the wave equation for the membrane

$$\frac{\partial^2\psi}{\partial x^2} + \frac{\partial^2\psi}{\partial y^2} = \frac{1}{c^2}\frac{\partial^2\psi}{\partial t^2} \tag{5.3}$$

The operator $(\partial^2/\partial x^2) + (\partial^2/\partial y^2)$ which operates on the wave displacement is the two-dimensional Laplace operator which we call ∇^2 and which some authors denote by the symbol Δ. The same symbol is used for three dimensions and the number two in ∇^2 is for the second order partial differential equation, the wave equation

$$\nabla^2\psi = \frac{1}{c^2}\frac{\partial^2\psi}{\partial t^2} \tag{5.4}$$

We rely on the physical picture shown in Fig. 5.1b that the solution to the wave equation may be the product of the motion in the x-direction with that of the y-direction and with the time variation to give

$$\psi = X(x)\,Y(y)\,e^{-i\omega t} \tag{5.5}$$

Differentiation and substitution into Eq. 5.3 gives

$$Y(y)\frac{\partial^2 X(x)}{\partial x^2} + X(x)\frac{\partial^2 Y(y)}{\partial y^2} = X(x)Y(y)\frac{\omega^2}{c^2} \tag{5.6}$$

where the common term $e^{-i\omega t}$ has been cancelled out. Collecting terms in x only on the left-hand side of this last equation gives

$$\frac{1}{X(x)}\frac{\partial^2 X(x)}{\partial x^2} = -\left(\frac{\omega^2}{c^2} + \frac{1}{Y(y)}\frac{\partial^2 Y}{\partial y^2}\right) \tag{5.7}$$

For the left side to be a function of x only and the right side to be a function of y only (aside from the constant $(\omega/c)^2$), each side must be equal to a constant, $-k^2$.

Thus we have two simple equations

$$\frac{\partial^2 X(x)}{\partial x^2} = -k^2 X(x) \tag{5.8a}$$

and

$$\frac{\partial^2 Y(y)}{\partial y^2} = +\left(k^2 - \frac{\omega^2}{c^2}\right) Y(y) \tag{5.8b}$$

which represent the separate space dependent parts of the displacement ψ.

The solution to Eq. 5.8a can be taken to be

$$X(x) = Ce^{\gamma x} \tag{5.9}$$

where both C and γ may be complex.

Differentiation and substitution into Eq. 5.8a gives

$$\gamma^2 = -k^2 \tag{5.10a}$$

or

$$\gamma = \pm ik \tag{5.10b}$$

and

$$X(x) = C_+ e^{ikx} + C_- e^{-ikx} \tag{5.11}$$

The space-dependent part in the x-direction resembles the flexible string with a wave running to the right and one to the left.

The solution to Eq. 5.8b is taken to be

$$Y(y) = De^{\alpha y} \tag{5.12}$$

Differentiation and substitution into Eq. 5.8b give

$$\alpha = k^2 - \frac{\omega^2}{c^2} \tag{5.13a}$$

or

$$\alpha = \pm i\sqrt{(\omega^2/c^2) - k^2} \tag{5.13b}$$

and

$$Y(y) = D_+ \exp i\left(\frac{\omega^2}{c^2} - k^2\right)^{1/2} y + D_- \exp -i\left(\frac{\omega^2}{c^2} - k^2\right)^{1/2} y \tag{5.14}$$

The wave vector for the y-direction running waves, $(\omega^2/c^2 - k^2)^{1/2}$, depends upon the values of ω and of the constant k. At this point we can write down the displacement as

$$\psi(x, y, t) = [C_+e^{ikx} + C_-e^{-ikx}]\left[D_+ \exp i\left(\frac{\omega^2}{c^2} - k^2\right)^{1/2} y + D_- \exp -i\left(\frac{\omega^2}{c^2} - k^2\right)^{1/2} y\right]e^{-i\omega t} \tag{5.15}$$

The boundary conditions require that the reflected waves in both the x-direction and the y-direction are perfectly reflected so that for a membrane of length, a, in the x-direction and of length, b, in the y-direction we may now write with a real amplitude, A, the equation

$$\psi = A \sin kx \sin\left(\frac{\omega^2}{c^2} - k^2\right)^{1/2} y \cos(\omega t - \phi) \tag{5.16}$$

The boundary conditions ($\psi = 0$) at $x = a$ and at $y = b$ require that

$$k = \frac{\pi m}{a} \quad \text{and} \quad \left(\frac{\omega^2}{c^2} - k^2\right)^{1/2} = \frac{\pi n}{b} \qquad (5.17a)\ (5.17b)$$

where m and n are integers. The phase angle, ϕ, is appropriate for the mode of motion determined by the integers m, n just as the amplitude is appropriate. Finally, we obtain a wave displacement characterized by these integers

$$\psi_{mn} = A_{mn} \sin\frac{\pi mx}{a} \sin\frac{\pi ny}{b} \cos(\omega_{mn}t - \phi_{mn}) \tag{5.18}$$

with

$$\omega = c\pi\sqrt{(m^2/a^2) + (n^2/b^2)} \equiv \omega_{mn} \tag{5.19}$$

the characteristic frequency.

The rectangular membrane has discrete allowed frequencies which may be calculated from Eq. 5.19. This requires a knowledge of the wave velocity, shape of the membrane, and appropriate integers m, n for the characteristic mode of motion. In the event the membrane is square shaped so that distance $a = b$, two modes of motion which look different to the eye as to physical shape will have an identical frequency. For example, a square membrane can have a mode of motion with the left-hand side up and the right side down as in Fig. 5.1a. This same square membrane could have a mode of motion with the top half up and the bottom half down, and thus a nodal line across the middle from left to right. These two modes of motion, characterized by $m = 2, n = 1$ in the first case and by $m = 1, n = 2$ in the second case, have the same frequency $\omega_{21} = \omega_{12}$ because the distance $a = b$. Modes of motion having the same frequency are called degenerate. In describing the modes of motion of electrons about the nucleus of an atom, it is possible to have quite different looking modes with the same frequency and this too is a condition of degeneracy. In such cases the degeneracy may be removed by an external magnetic field (Zeeman splitting), and this is an important part of the physics of magnetism.

The shape of a membrane may be described by use of a Fourier series of the characteristic modes we have just discussed and represented by Eq.5.18. We may write

$$\psi = \sum_{m=1}^{\infty} \sum_{n=1}^{\infty} A_{mn} \sin \frac{\pi m x}{a} \sin \frac{\pi n y}{b} \cos (\omega_{mn} t - \phi_{mn}) \qquad (5.20a)$$

or

$$\psi = \sum_{m=1}^{\infty} \sum_{n=1}^{\infty} \sin \frac{\pi m x}{a} \sin \frac{\pi n y}{b} [B_{mn} \cos (\omega_{mn} t) + C_{mn} \sin \omega_{mn} t)] \qquad (5.20b)$$

The coefficients may be calculated from a knowledge of the initial shape or initial velocity, just as the Fourier coefficients were calculated in Chapter 2, Eqs. 2.18 etc.

5.2 THE FLEXIBLE MEMBRANE WITH CIRCULAR BOUNDARY: Bessel Functions

The observable properties of thin flexible membranes with circular boundaries must be imagined because even the familiar kettledrum is not so ideal and is stretched over an airtight vessel. But the keetledrum does allow us to observe that a normal or characteristic mode is a function of the mass per unit area, tension per unit length of circumference, and that nodal lines can be observed as concentric rings as shown in Fig. 5.2. Nodal lines may also extend across a diameter. The great value of our study of the mathematical tools required for the circular membrance treated in polar coordinates is that we learn of some new characteristic functions—Bessel functions. These are very special Bessel functions and our discussion is not meant to replace the general presentation found in a standard text on Advanced Calculus. We hope that the present discussion will enrich and deepen the reader's interest in the others.

The segment of area in polar coordinates is shown in Fig. 5.2*b* and the component of force in the vertical or displacement direction is the sum of two parts:

$$(1) \quad T\,dr \left[\left(\frac{\partial \psi}{r d\phi} \right)_{\phi + d\phi} - \left(\frac{\partial \psi}{r \partial \phi} \right)_{\phi} \right] = T \frac{\partial^2 \psi}{r \partial \phi^2} d\phi\, dr = \frac{T}{r^2} \frac{\partial^2 \psi}{\partial \phi^2} r\, d\phi\, dr \qquad (5.21a)$$

$$(2) \quad T\,d\phi \left[\left(r \frac{\partial \psi}{\partial r} \right)_{r+dr} - \left(r \frac{\partial \psi}{\partial r} \right)_{r} \right] = \frac{T}{r} \frac{\partial}{\partial r} \left(r \frac{\partial \psi}{\partial r} \right) r\, d\phi\, dr \qquad (5.21b)$$

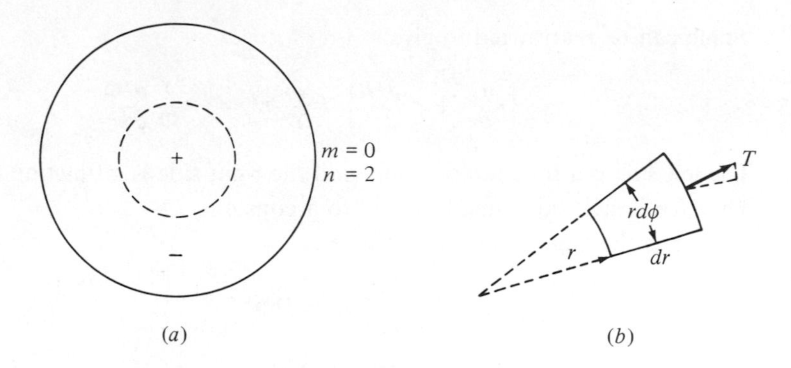

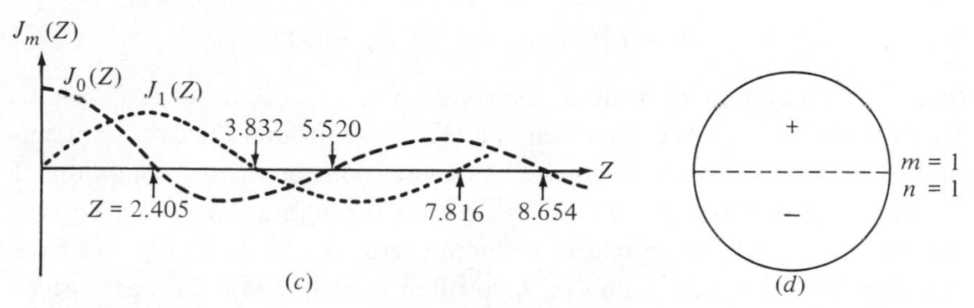

Figure 5.2a Circular Membrane. **b** Segment of Circular Membrane. **c** Characteristic Radial Wave Function for Membrane, Bessel Functions. **d** Characteristic Mode of Motion on Circular Membrane.

The sum of these two force components must equal the mass times the vertical acceleration. Thus

$$T\left\{\frac{1}{r^2}\frac{\partial^2\psi}{\partial\phi^2} + \frac{1}{r}\frac{\partial}{\partial r}\left(r\frac{\partial\psi}{\partial r}\right)\right\} r\,d\phi\,dr = \rho r\,d\phi\,dr\frac{\partial^2\psi}{\partial t^2} \tag{5.22}$$

Cancelling common terms gives the wave equation

$$\frac{1}{r^2}\frac{\partial^2\psi}{\partial\phi^2} + \frac{1}{r}\frac{\partial}{\partial r}\left(r\frac{\partial\psi}{\partial r}\right) = \frac{\rho}{T}\frac{\partial^2\psi}{\partial t^2} = \frac{1}{c^2}\frac{\partial^2\psi}{\partial t^2} \tag{5.23}$$

The solution we try in order to separate variables is

$$\psi = R(r)\Phi(\phi)e^{-i\omega t} \tag{5.24}$$

Differentiation and substitution into the wave equation gives

$$\frac{1}{r}\left\{\frac{\partial R}{\partial r}\Phi + r\frac{\partial^2 R}{\partial r^2}\Phi\right\} + \frac{1}{r^2}R\frac{\partial^2\Phi}{\partial\phi^2} = -\frac{\omega^2}{c^2}R\Phi \tag{5.25a}$$

which can be rearranged to give

$$\frac{1}{R}\left\{r\frac{\partial R}{\partial r} + r^2\frac{\partial^2 R}{\partial r^2}\right\} + \frac{\omega^2}{c^2}r^2 = -\frac{1}{\Phi}\frac{\partial^2 \Phi}{\partial \phi^2} \tag{5.25b}$$

The left side is a function of r only and the right side is a function of ϕ only. Therefore, each side must be equal to a constant, l^2.

Thus the angular dependent part is

$$\frac{\partial^2 \Phi}{\partial \phi^2} = -l^2\Phi \tag{5.26}$$

and this familiar equation has the solution

$$\Phi = e^{+il\phi} + e^{-il\phi} = \cos l(\phi - \alpha) \tag{5.27}$$

or a sine function could be used as a solution.

The solution gives a wave running clockwise around the circular membrane and a wave running counterclockwise. The boundary condition of the circular wave fitting onto itself as ϕ runs through an angle 2π requires that the constant, l, be an integer including zero; ($l = 0, 1, 2, \ldots$). We have just seen the l quantum numbers, l, so often used in quantum wave mechanics and its appearance will be a familiar friend.

The radial dependent part of the wave is also set equal to the constant l^2 and gives

$$r^2\frac{\partial^2 R}{\partial r^2} + r\frac{\partial R}{\partial r} + \left(\frac{\omega^2}{c^2}r^2 - l^2\right)R = 0 \tag{5.28}$$

Let $Z = (\omega/c)r$ and substitute into the above equation to obtain

$$Z^2\frac{\partial^2 R}{\partial Z^2} + Z\frac{\partial R}{\partial Z} + (Z^2 - l^2)R = 0 \tag{5.29}$$

Try the power series solution

$$R = a_0Z^m + a_1Z^{m+1} + a_2Z^{m+2} + \cdots a_rZ^{m+r} + \cdots \tag{5.30}$$

Differentiation and equating to zero the sum of the coefficients of each of the like powers of Z gives

$$(m^2 - l^2)a_0 = 0 \qquad \text{so} \qquad m = \pm l \tag{5.31a}$$

$$[(m+1)^2 - l^2]a_1 = 0 \qquad \text{so} \qquad a_1 = 0 \quad \text{if } m = +l \tag{5.31b}$$

$$[(m+2)^2 - l^2]a_2 + a_0 = 0 \qquad \text{so} \qquad a_2 = \frac{a_0}{2(2l+2)} \tag{5.31c}$$

$$[(m+r)^2 - l^2]a_r + a_{r-2} = 0 \qquad \text{so} \qquad a_r = -\frac{a_{r-2}}{r(2l+r)} \tag{5.31d}$$

Thus for $m = +l$ we have the series:

$$R = a_0 Z^m \left(1 - \frac{Z^2}{2(2m+2)} + \frac{Z^4}{2 \cdot 4(2m+2)(2m+4)} - \frac{Z^6}{2 \cdot 4 \cdot 6(2m+2)(2m+4)(2m+6)} + \cdots\right) \tag{5.32}$$

But for $m = -l$ we would get

$$R = a_0 Z^{-l} \left(1 + \frac{Z^2}{2(2l-2)} + \frac{Z^4}{2 \cdot 4(2l-2)(2l-4)} + \cdots\right) \tag{5.33}$$

With $Z = (\omega/c)r$ we have the series becoming infinite for $r = 0$ so we exclude this series as physically unacceptable. There will be problems similar to this one in later sections of this book in which such a second radial solution is not excluded because an inner radius boundary condition does not allow $r \longrightarrow 0$. The series for the allowed radial wave displacement can be easily identified with the Bessel function of the first kind of order m (for we may allow $a_0 = 1/2^m\, m!$):

$$J_m(Z) = \frac{Z^m}{2^m m!}\left(1 - \frac{Z^2}{2^2(m+1)} + \frac{Z^4}{2^4 2!(m+1)(m+2)} - \frac{Z^6}{2^6 3!(m+1)(m+2)(m+3)} + \cdots\right) \tag{5.34a}$$

$$= \frac{1}{m!}\left(\frac{Z}{2}\right)^m - \frac{1}{(m+1)!}\left(\frac{Z}{2}\right)^{m+2} + \frac{1}{2!(m+2)!}\left(\frac{Z}{2}\right)^{m+4} + \cdots \tag{5.34b}$$

In particular we have the series shown in Fig. 5.2c:

$$J_0(Z) = 1 - \frac{Z^2}{2^2} + \frac{Z^4}{2^4(2!)^2} + \cdots \tag{5.35}$$

$$J_1(Z) = \frac{Z}{2} - \frac{Z^3}{2^3 2!} + \frac{Z^5}{2^5 2! 3!} + \cdots \tag{5.36}$$

We can imagine a certain similarity with cos (Z) and sin (Z) functions whose amplitude diminishes as Z becomes large, but the Bessel functions are not strictly periodic. We can easily see that these radial functions are ideal for describing waves that spread out on a membrane and lower their amplitude as $Z = (\omega/c)r$ becomes large. An easy check will show that

$$\frac{d}{dZ} J_0(Z) = -J_1(Z) \tag{5.37}$$

which is again a similarity with cos (Z) and sin (Z) functions. Returning to our circular membrane, the solution to the wave equation for a given integer value of m is at this point

$$\psi_m = A_m \cos m(\phi - \alpha) J_m\left(\frac{\omega r}{c}\right) \cos (\omega t - r) \tag{5.38}$$

We have now the restriction at the radial boundary that

$$\psi = 0 \qquad \text{when } r = a$$

The Bessel functions do have zeros along the axis $Z = \omega r/c$ and thus a frequency ω_{mn} is allowed such that $J_m(\omega_{mn}a/c) = 0$. If $n = 1$, the first zero of the Bessel function is at the boundary. If $n = 2$, the second zero is at the circumference of the membrane. The wave on the circular membrane oscillating in the characteristic mode determined by the integers m and n is given as

$$\psi_{mn} = A_{mn} \begin{Bmatrix} \sin \\ \cos \end{Bmatrix} (m\phi) J_m\left(\frac{\omega_{mn} r}{c}\right) e^{-i\omega_{mn}t} \tag{5.39}$$

The integer m gives the number of wavelengths in the azimuthal angle 2π. The integer n gives the number of half wavelengths in the distance a along the radius. A Fourier series representation of some displacement of the membrane may be made up with these new characteristic functions and the motion followed as a function of time. In order to apply the Fourier series method to a useful and practical example, we turn next to the forced motion of the rectangular and of the circular membrane. We shall discuss these two cases side by side.[1]

5.3 FORCED MOTION ON MEMBRANES

When a driving force is applied to a membrane, an extra term must be added to the equation of motion.

$$\underbrace{T\nabla^2\eta}_{\substack{\text{restoring force}\\ \text{per unit area}}} + \underbrace{F}_{\substack{\text{applied force}\\ \text{per unit area}}} = \underbrace{\sigma\frac{\partial^2\eta}{\partial t^2}}_{\substack{\text{(mass/unit area)}\cdot\\ \text{acceleration}}} \tag{5.40}$$

In general, this applied force per unit area (i.e., pressure) varies both with

[1]The author wishes to acknowledge that the following paragraphs about forced motion on membranes are similar to a set of classroom notes by an unknown instructor written during the World War II years 1941-45 and not signed or copyrighted. The material has been so well received over the years by students that we express our thanks for a well-done teaching assist.

time and with position on the surface of the membrane. For simplicity we consider the variation with time to be sinusoidal and represent the applied pressure by

$$F = P(x, y)e^{-i\omega t} \qquad (5.41a)$$

$$F = P(r, \phi)e^{-i\omega t} \qquad (5.41b)$$

where ω is the angular frequency of the applied pressure and the quantities $P(x, y)$ and $P(r, \phi)$ describe the space variation of the pressure over the surface of the membrane.

Suppose the space variation of the applied pressure has the same form as the space variation corresponding to one of the natural frequencies of the membrane, but the frequency of the applied force *differs* from the natural frequency. For example, let the force per unit area be

$$F = \underbrace{P_{12}}_{\substack{\text{amplitude}\\ \text{factor of}\\ \text{pressure}}} \underbrace{\sin \frac{\pi x}{a} \sin \frac{2\pi y}{b}}_{\substack{\text{space variation}\\ \text{of pressure}}} \underbrace{e^{-i\omega t}}_{\substack{\text{time}\\ \text{variation}\\ \text{of pressure}}} \qquad (5.42a)$$

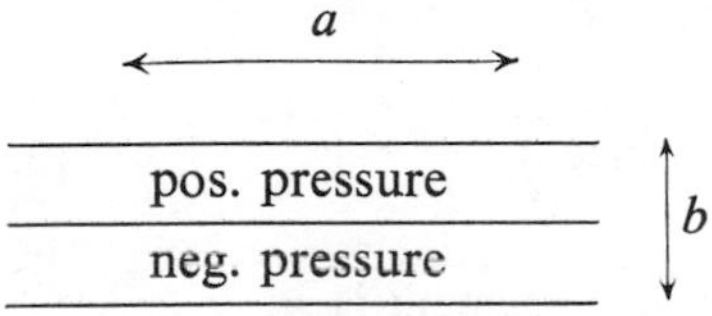

a

pos. pressure

neg. pressure

b

$m = 1,\ n = 2$

$$F = \underbrace{P_{11}}_{\substack{\text{amplitude}\\ \text{factor of}\\ \text{pressure}}} \underbrace{\cos(\phi) J_1\left(\frac{\pi \beta_{11} r}{a}\right)}_{\substack{\text{space variation}\\ \text{of pressure}}} \underbrace{e^{-i\omega t}}_{\substack{\text{time}\\ \text{variation}\\ \text{of pressure}}} \qquad (5.42b)$$

where

$$\nu_{mn} = \frac{c}{2a}\beta_{mn}$$

$-$ $+$

$m = 1,\ n = 1$

The *free* vibration displacement with the corresponding space variation was shown to be:

$$\eta_{12} = A_{12} \sin \frac{\pi x}{a} \sin \frac{2\pi y}{b} e^{-i\omega t} \qquad (5.43a)$$

$$\eta_{11} = A_{11} \cos(\phi) J_1\left(\frac{\pi \beta_{11} r}{a}\right) e^{-i\omega t} \qquad (5.43b)$$

The *forced* displacement must have the same frequency as the driving force. Since the driving force is pushing the membrane in a natural pattern, i.e., in a pattern the membrane would choose if free, the forced displacement assumes the same space pattern as the free vibration, but has the frequency of the applied force. We therefore *assume* that the forced displacement is

$$\eta - A_{12} \sin \frac{\pi x}{a} \sin \frac{2\pi y}{b} e^{-i\omega t} \qquad (5.44a)$$

$$\eta = A_{11} \cos(\phi) J_1\left(\frac{\pi \beta_{11} r}{a}\right) e^{-i\omega t} \qquad (5.44b)$$

To determine the amplitude of the displacement, the assumed expression for displacement is substituted into the wave equation. For rectangular coordinates:

$$\frac{\partial^2 \eta}{\partial x^2} + \frac{\partial^2 \eta}{\partial y^2} + \frac{P_{12}}{T} \sin \frac{\pi x}{a} \sin \frac{2\pi y}{b} e^{-i\omega t} - \frac{1}{c^2} \frac{\partial^2 \eta}{\partial t^2} = 0 \tag{5.45}$$

or

$$A_{12}\left[\left(-\frac{\pi^2}{a^2}\right) + \left(-\frac{4\pi^2}{b^2}\right)\right] \sin \frac{\pi x}{a} \sin \frac{2\pi y}{b} e^{-i\omega t}$$

$$+ \frac{P_{12}}{T} \sin \frac{\pi x}{a} \sin \frac{2\pi y}{b} e^{-i\omega t}$$

$$+ \frac{A_{12}}{c^2} \omega^2 \sin \frac{\pi x}{a} \sin \frac{2\pi y}{b} e^{-i\omega t} = 0 \tag{5.46}$$

Cancelling the common factors gives

$$A_{12}\left[\left(\frac{\pi^2}{a^2} + \frac{4\pi^2}{b^2}\right) - \frac{\omega^2}{c^2}\right] = \frac{P_{12}}{T} \tag{5.47a}$$

or

$$A_{12}\left[\frac{\omega_{12}^2}{c^2} - \frac{\omega^2}{c^2}\right] = \frac{P_{12}}{T} \tag{5.47b}$$

So

$$A_{12} = \frac{P_{12}}{\sigma(\omega_{12}^2 - \omega^2)} \qquad \text{since } \frac{T}{c^2} = \sigma \tag{5.48}$$

and the complete expression for the forced displacement is

$$\eta = \underbrace{\frac{P_{12}}{\sigma(\omega_{12}^2 - \omega^2)}}_{\text{amplitude factor}} \underbrace{\sin \frac{\pi x}{a} \sin \frac{2\pi y}{b}}_{\text{space variation}} \underbrace{e^{-i\omega t}}_{\text{time variation}} \tag{5.49}$$

amplitude factor showing resonance

It can be seen that the amplitude of the forced displacement has a resonance when the frequency of the driving force equals the natural frequency for the space pattern of the displacement.

For polar coordinates:

$$\frac{1}{r} \frac{\partial}{\partial r}\left(r \frac{\partial \eta}{\partial r}\right) + \frac{1}{r^2} \frac{\partial^2 \eta}{\partial \phi^2} - \frac{1}{c^2} \frac{\partial^2 \eta}{\partial t^2} + \frac{P_{11}}{T} \cos \phi J_1\left(\frac{\pi \beta_{11} r}{a}\right) e^{-i\omega t} = 0 \tag{5.50}$$

assume

$$\eta = A_{11} \cos(\phi) J_1\left(\frac{\pi \beta_{11} r}{a}\right) e^{-i\omega t} \tag{5.51}$$

But, by definition, $J_1(z)$ satisfies the equation:

$$\frac{1}{z}\frac{d}{dz}\left(z\frac{dJ_1(z)}{dz}\right) + \left(1 - \frac{1}{z^2}\right)J_1(z) = 0 \tag{5.52}$$

So since $z = \pi\beta_{11}r/a$ we have

$$\frac{1}{r}\frac{\partial}{\partial r}\left(r\frac{\partial n}{\partial r}\right) = -A_{11}\cos\phi \cdot e^{-i\omega t} \cdot \left(\frac{\pi\beta_{11}}{a}\right)^2\left[1 - \left(\frac{a}{\pi\beta_{11}r}\right)^2\right]J_1\left(\frac{\pi\beta_{11}r}{a}\right) \tag{5.53a}$$

$$= A_{11}\left[-\frac{\pi^2\beta_{11}^2}{a^2} + \frac{1}{r^2}\right]\cos\phi \cdot J_1\left(\frac{\pi\beta_{11}r}{a}\right)e^{-i\omega t} \tag{5.53b}$$

By differentiation

$$\frac{1}{r^2}\frac{\partial^2\eta}{\partial\phi^2} = -\frac{A_{11}}{r^2}\cos\phi J_1\left(\frac{\pi\beta_{11}r}{a}\right)e^{-i\omega t} \tag{5.54a}$$

and

$$-\frac{1}{c^2}\frac{\partial^2\eta}{\partial t^2} = +A_{11}\frac{\omega^2}{c^2}\cos\phi J_1\left(\frac{\pi\beta_{11}r}{a}\right)e^{-i\omega t} \tag{5.54b}$$

So, putting these values into the wave equation and cancelling common factors, we get

$$A_{11}\left[-\frac{\pi^2\beta_{11}^2}{a^2} + \frac{\omega^2}{c^2}\right] + \frac{P_{11}}{T} = 0 \tag{5.55a}$$

So since $\quad \dfrac{\pi\beta_{11}}{a} = \dfrac{\omega_{11}}{c} \quad$ and $\quad \dfrac{T}{c^2} = \sigma$

$$A_{11} = \frac{P_{11}}{(\omega_{11}^2 - \omega^2)\sigma} \tag{5.55b}$$

and the complete displacement is given by

$$\eta = \frac{P_{11}}{\sigma(\omega_{11}^2 - \omega^2)}\cos(\phi)J_1\left(\frac{\pi\beta_{11}r}{a}\right)e^{-i\omega t} \tag{5.56}$$

Hence, in both rectangular coordinates and polar coordinates, a driving force having a space pattern corresponding to one of the normal modes of the membrane produces a displacement having the same space pattern with an amplitude which goes through a resonance when the driving frequency equals the natural frequency.[2]

But an applied force having a space pattern corresponding exactly to

[2]An interesting experiment on exactly this theoretical example has been published by I. Rudnick, H. Kojima, W. Veith, and R. S. Kagiwada, *Phys. Rev. Letters*, **23**, 1220 (1969).

one of the natural displacement patterns of the membrane is a very special case. When the space pattern of the applied force differs from the natural space patterns of the displacement, the solution is more complicated, but it can still be carried out by analyzing the space pattern of the actual applied force in terms of components which do have the same pattern of space variation as the natural modes of displacement of the membrane for free vibrations. By a process similar to Fourier analysis, almost any space distribution of pressure can be resolved into components of this type. For example, suppose the applied pressure is a constant over the whole surface of the membrane. This is a simple and common case, but this space distribution does not correspond to any of the displacement patterns for the free vibrations of the membrane. In this case, the force/unit-area is given by a constant times the sinusoidal time factor

$$F = \underbrace{P}_{\text{const.}} \underbrace{e^{-i\omega t}}_{\substack{\text{sinusoidal} \\ \text{time variation}}} \tag{5.57}$$

We assume that this force can be resolved into component forces which have different amplitudes and have space distributions which correspond to the patterns of the various natural modes of free vibration

$$P = \sum_{mn} P_{mn} \sin \frac{m\pi x}{a} \sin \frac{n\pi y}{b} \tag{5.58a}$$

$$P = \sum_{n} P_{0n} \cdot 1 \cdot J_0\left(\frac{\pi \beta_{0n} r}{a}\right) \tag{5.58b}$$

Since the pressure is uniform and hence symmetric, the asymmetric functions (i.e., with $m > 0$) are not needed here.

The amplitude of each component, P_{mn}, is determined by making use of the integral properties of the functions which describe the space patterns of the displacement in free vibration.

$$\int_0^a \sin \frac{m\pi x}{a} \sin \frac{k\pi x}{a} dx = \begin{cases} \frac{1}{2}(a) & \text{if } m = k \\ 0 & \text{if } m \neq k \end{cases}$$

$$\int_0^b \sin \frac{n\pi y}{b} \sin \frac{l\pi y}{b} dy = \begin{cases} \frac{1}{2}(b) & \text{if } n = l \\ 0 & \text{if } n \neq l \end{cases} \tag{5.59a}$$

$$\int_0^{2\pi} \int_0^a J_0\left(\frac{\pi \beta_{0n} r}{a}\right) J_0\left(\frac{\pi \beta_{0l} r}{a}\right) r\, dr\, d\phi = \begin{cases} \pi a^2 [J_1(\pi \beta_{0n})]^2 & \text{if } n = l \\ 0 & \text{if } n \neq l \end{cases} \tag{5.59b}$$

So the value of each amplitude factor, P_{kl}, can be calculated by the relations obtained by multiplying the const. P by

$$\sin\frac{k\pi x}{a}\sin\frac{l\pi y}{b}\,dx\,dy$$

$$J_0\left(\frac{\pi\beta_{0l}r}{a}\right)r\,dr\,d\phi$$

since $k = 0$

and integrating. The resulting relations are:

$$\int_0^a\int_0^b P\cdot\sin\frac{k\pi x}{a}\sin\frac{l\pi y}{b}\,dx\,dy = \tfrac{1}{4}(ab)P_{kl} \qquad (5.60a)$$

$$2\pi\int_0^a P\cdot rJ_0\left(\frac{\pi\beta_{0l}r}{a}\right)dr = \pi a^2(J_1(\pi\beta_{0l}))^2P_{0l} \qquad (5.60b)$$

In the present case P is a constant and can come outside the integral, giving:

$$P\left[\frac{a}{k\pi}\cos\frac{k\pi x}{a}\right]_0^a\cdot\left[\frac{b}{l\pi}\cos\frac{l\pi y}{b}\right]_0^b = \tfrac{1}{4}abP_{kl} \qquad (5.61a)$$

$$2P\int_0^a r\cdot J_0\left(\frac{\pi\beta_{0l}r}{a}\right)dr = a^2(J_1(\pi\beta_{0l})^2)P_{0l} \qquad (5.61b)$$

But $\cos k\pi = \begin{cases} 1 \text{ if } k \text{ even} \\ -1 \text{ if } k \text{ odd}\end{cases}$

and $\cos 0 = 1$

So

$$P\frac{ab}{lk\pi^2}\cdot 4 = \frac{ab}{4}P_{kl} \qquad (5.62a)$$

where k and l are now restricted to odd values.

So

$$P_{kl} = \left(\frac{16}{lk\pi^2}\right)P \text{ for } k \text{ and } l \text{ odd}$$

$$= 0 \text{ for } k \text{ or } l \text{ even} \qquad (5.63)$$

But $\int zJ_0(z)\,dz = zJ_1(z)$

So

$$\frac{2a}{\pi\beta_{0l}}P\left[rJ_1\left(\frac{\pi\beta_{0l}r}{a}\right)\right]_0^a = a^2(J_1(\pi\beta_{0l}))^2P_{0l} \qquad (5.62b)$$

β_{0l} and $J_1(\pi\beta_{0l})$ are tabulated numbers which vary only with l. See Fig. 5.2c where the value $\pi\beta_{0l}$ makes $J_0(\pi\beta_{0l}) = 0$ and thus $\pi\beta_{01} = 2.405$, $\pi\beta_{02} = 5.520$, etc.

$\beta_{01} = 0.7655 \quad J_1(\pi\beta_{01}) = 0.5191$

$\beta_{02} = 1.7571 \quad J_1(\pi\beta_{02}) = -0.3403$

$\beta_{03} = 2.7546 \quad J_1(\pi\beta_{03}) = 0.2715$

Calculated from Eq. 5.36

We have broken the uniform pressure distribution into a sum of components and have determined the amplitude of each component (P_{kl}). Each of these components has a space distribution of pressure identical with the amplitude distribution of some one of the normal modes of free vibration. As we saw above, any one of these pressure components excites only the corresponding normal mode of vibration. Furthermore, the amplitude of that normal mode is a maximum (resonance) when the driving frequency is equal to the normal frequency of that mode.

Now the total displacement amplitude is given by the sum of all of the component displacement amplitudes. So the general procedure for solving a problem involving forced motion is as follows:

Given: (1) a pressure with a known space distribution and a sinusoidal time variation.
(2) a membrane with a series of normal modes, each characterized by a space distribution of displacement (i.e., a characteristic function) and by a resonant frequency for free vibration.

First, resolve applied force into components having a space distribution identical with the space distribution of normal modes of displacement for free vibration.

$$\text{applied force} = F = P(x, y)e^{-i\omega t}$$

$$P(x, y) = \sum_{m,n} P_{mn} \sin \frac{m\pi x}{a} \sin \frac{n\pi y}{b} \tag{5.64}$$

where

$$P_{mn} = \frac{4}{ab}\int_0^a \int_0^b P(x, y) \sin \frac{m\pi y}{a} \sin \frac{n\pi y}{b}\, dx\, dy \tag{5.65}$$

Second, solve for the displacement due to each component force independently. This gives component displacements.

$$\eta_{mn} = \frac{P_{mn}}{\sigma(\omega_{mn}^2 - \omega^2)} \sin \frac{m\pi x}{a} \sin \frac{n\pi y}{b} e^{-i\omega t} \tag{5.66}$$

Third, add all component displacements to get the total displacement due to total applied force.

$$\eta = \sum_{m,n} \frac{P_{mn}}{\sigma(\omega_{mn}^2 - \omega^2)} \sin \frac{m\pi x}{a} \sin \frac{n\pi y}{b} e^{-i\omega t} \tag{5.67a}$$

$$= \sum_{mn} \eta_{mn} \tag{5.67b}$$

PROBLEMS AND SOLUTIONS TO PROBLEMS

1. Consider a square flexible membrane 4 cm on a side with density 10^{-2} gram/cm^2 and tension 10^6 dynes/cm. Let the membrane be driven by a force of $10^4 \cos \omega t$ dynes/cm^2 which is uniform across the membrane. What is the amplitude of displacement of the midpoint at frequencies of 400, 800, 1200, 1600, 1800 and 2000 cycles per second? There is no damping action on this ideal membrane.

Solution: The uniform-in-space-coordinates pressure is expanded in Fourier series as

$$P = \sum_m \sum_n P_{mn} \sin \frac{\pi m x}{a} \sin \frac{\pi n y}{a}$$

The Fourier coefficient is

$$\frac{a}{2} \cdot \frac{a}{2} P_{mn} = \int_0^a \int_0^a P \sin \frac{\pi m x}{a} \sin \frac{\pi n y}{a} dx\, dy$$

or

$$P_{mn} = \frac{4}{a^2} P \left[\frac{a}{m\pi} \cos \frac{m\pi x}{a} \right]_0^a \left[\frac{a}{n\pi} \cos \frac{n\pi y}{a} \right]_0^a$$

$$P_{mn} = \frac{16P}{mn\pi^2}$$

where both m and n are odd integers.

Thus the displacement at the midpoint is

$$|\eta| = \frac{16 \times 10^6}{4x\pi^4} \sum_m \sum_n \frac{\sin m(\pi/2) \sin n(\pi/2)}{mn(\nu_{mn}^2 - \nu^2)}$$

where

$$\nu_{mn} = \frac{1}{2} \sqrt{\frac{T}{\sigma}} \frac{\sqrt{m^2 + n^2}}{a} = \frac{10^4}{2 \times 4} \sqrt{m^2 + n^2}$$

and the normal mode frequencies are

$$\nu_{11} = 1770 \text{ cps}, \nu_{13} = \nu_{31} = 3950 \text{ cps}, \nu_{33} = 5300 \text{ cps, etc.}$$

Writing the first three terms (including the double degeneracy) gives

$$|\eta(2, 2)| = \frac{1.62 \times 10^6}{4\pi^2} \left[\frac{1}{\nu_{11}^2 - \nu^2} - \frac{2}{3(\nu_{13}^2 - \nu^2)} + \frac{1}{9(\nu_{33}^2 - \nu^2)} \right]$$

or $$|\eta| = 4.1 \times 10^4 \left[\frac{1}{(1770)^2 - \nu^2} - \frac{2}{3[(3950)^2 - \nu^2]} + \frac{1}{9[(5300)^2 - \nu^2]} \right]$$

Obviously, more terms could be taken but they are quite small. Thus the amplitude at the midpoint is computed to be

ν	$\lvert\eta\rvert$ cm
400	1.22×10^{-2}
800	1.48×10^{-2}
1200	2.25×10^{-2}
1600	7.07×10^{-2}
1770	∞
1800	37.45×10^{-2}
2000	4.93×10^{-2}

2. A flexible, undamped, circular membrane of radius 2 cm, density 0.1 gm/cm², tension 6.31×10^5 dynes/cm is driven by a uniform force of $4 \times 10^5 \cos \omega t$ dynes/cm² across the membrane. What is the amplitude of the midpoint at frequencies of 200, 400, 600, 800, 1000, and 1200 cps? The solution is as follows: We Fourier expand the uniform in space coordinates pressure as

$$P = \sum_n P_{0n} J_0\left(\frac{\pi \beta_{0n} r}{a}\right)$$

We calculate the Fourier coefficients by writing

$$P \cdot J_0\left(\frac{\pi \beta_{0l} r}{a}\right) = \sum_n P_{0n} J_0\left(\frac{\pi \beta_{0n} r}{a}\right) J_0\left(\frac{\pi \beta_{0l} r}{a}\right)$$

Integration gives

$$\int_0^a \int_0^{2\pi} P J_0\left(\frac{\pi \beta_{0l} r}{a}\right) r\, dr\, d\phi = \iint \sum_n P_{0n} J_0\left(\frac{\pi \beta_{0n} r}{a}\right) J_0\left(\frac{\pi \beta_{0l} r}{a}\right) r\, dr\, d\phi$$

or

$$2\pi P \int_0^a r J_0\left(\frac{\pi \beta_0 r}{a}\right) dr = \pi a^2 [J_1(\pi \beta_{0l})]^2 P_{0l}$$

and

$$P_{0l} = \frac{2P}{\pi \beta_{0l}[J_1(\pi \beta_{0l})]}$$

The displacement is

$$\eta = \sum_l \frac{P_{0l}}{\sigma(\omega_{0l}^2 - \omega^2)} J_0\left(\frac{\pi \beta_{0l} r}{a}\right) \cos \omega t$$

At the center $r = 0$ so that

$$J_0\left(\frac{\pi \beta_{0l} r}{a}\right) = 1$$

The displacement amplitude is

$$|\eta| = \frac{2P}{\sigma} \sum_l \frac{1}{\pi \beta_{0l}[J_1(\pi \beta_{0l})][\omega_{0l}^2 - \omega^2]}$$

or

$$|\eta| = \frac{2P}{4\pi^2 \sigma} \sum \frac{1}{\pi \beta_{0l}[J_1(\pi \beta_{0l})][\nu_{0l}^2 - \nu^2]}$$

The normal mode frequencies are

$$\nu_{0n} = \frac{c}{2a} \beta_{0n}, \qquad \nu_{01} = 481,\ \nu_{02} = 1{,}105,\ \nu_{03} = 1{,}735 \text{ cps}$$

Using these first three normal modes, we have

$$|\eta| = 2.02 \times 10^5 \left[\frac{1}{1.25(\nu_{01}^2 - \nu^2)} - \frac{1}{1.88(\nu_{02}^2 - \nu^2)} + \frac{1}{2.35(\nu_{03}^2 - \nu^2)}\right]$$

Thus we calculate

ν cps	$\lvert \eta \rvert$ cm
200	7.83×10^{-1}
400	22.2×10^{-1}
481	∞
600	13.5×10^{-1}
800	5.45×10^{-1}
1000	6.55×10^{-1}
1105	∞
1200	4.11×10^{-1}

It is instructive to calculate the entire space dependence at some fixed driving frequency in order to use the Bessel function $J_0(x)$.

Chapter **6**

PLANE WAVE PROPAGATION OF SOUND

The present chapter introduces the basic ideas of three-dimensional wave propagation by pressure waves in the medium. The theory of sound waves, including the theory of vibrating strings and membranes which we have reviewed in the earlier chapters of this book, was written in excellent detail as a textbook by Lord Rayleigh in 1877. Indeed many texts on sound have been written during the past century which owe their basic content to the writing of Lord Rayleigh and others. The justification for distilling down such great works is the ever present need to give the on-coming gneration of scientists and engineers the essential concepts and to make room for new approaches. But we urge the reader to enjoy the writings of the great 19th century Nobel Laureate, John William Strutt, Baron Rayleigh: *The Theory of Sound.*[1] For example, Rayleigh's account of underwater sound measurements in 1826 by Colladon and Sturm in the Lake of Geneva is one of the many historical gems. A bell under water was struck at the same instant that the flash of a cannon could be seen from a distant point of the lake. A tube extending down into the water at that distant point finally brought the underwater sound of the struck bell, and from the time lapse and known distance a very good value for underwater sound velocity was determined. These early workers also noted that the higher frequencies were strongly attenuated. By contrast, the photograph showing the sound dome on the USS *Willis A. Lee* is a modern application of the wave propagation which we should understand.

[1] John William Strutt, Baron Rayleigh, *The Theory of Sound*, 2 vol., 2nd edition, revised and enlarged, Dover Publications, New York (1945).

Photograph of bow of the U.S.S. *Willis A. Lee,* showing underwater sound dome for ultrasonic echo ranging. Photo courtesy Naval Photographic Center, Washington, D.C.

The waves on strings and membranes discussed in the earlier chapters have been transverse motion of amplitude, $\psi(x, t)$. The physical properties of the medium involving density and tension were used to characterize the wave velocity and impedance of the transmission line. In the case of pressure waves in a medium, the physical properties of the medium must also be understood. The nature of the wave oscillation may then be described with precision. Sound waves in air, or in any gas, are compressional and dilatational modes of motion of the molecules and these are along the direction of wave propagation. Transverse motion is not given wave propagation because shear is not supported by the gas at standard pressure and temperature. The transverse waves can be transmitted by solids as well as longitudinal waves, and each wave type has generally quite different wave velocity. A liquid can easily transmit pressure waves which are longitudinal at all frequencies and can, to a very small extent, support transverse waves of very high frequency. This will remind us that a liquid is considered as somewhat gas-like and somewhat like a solid. To understand how vibrating devices can set up pressure waves in such media and how these waves propagate, we turn our attention to the physical properties of gases, liquids, and solids. Indeed our emphasis will be on the knowledge gained about the medium through careful measurement of pressure wave propagation.

The phase diagram shown in Fig. 6.1*a* is a schematic (not to scale) of a classical substance such as argon. The liquid and vapor are in equilibrium at any given temperature along the solid line separating the two phases. Likewise, the solid and vapor are separated along the sublimation curve. These join at the triple point where solid, liquid, and vapor are in equilibrium at a specific pressure and temperature for each substance. More complicated phase diagrams for more complicated molecules exist than the simple one illustrated here. For example, water expands on turning to ice and has a unique diagram in the region of the triple point. The plot of pressure vs. volume is shown in Fig. 6.1*b* and the dashed line is an isotherm. These familiar phase diagrams cannot yet be explained by starting with a knowledge of intermolecular forces and proceeding to a quantitative theory for condensation. We know the transition from gas to liquid is abrupt and the compressibility $K_T = -(1/V)(\partial V/\partial P)_T$ is infinite at the transition. We shall see that a pressure wave in a medium with infinite compressibility does not propagate. To avoid these regions we shall confine our discussion of the properties of matter to the pure gas state, or the pure liquid, or the pure solid state.

In Fig. 6.1*c* we show the spherically symmetric potential which may be used in discussing the attractive and repulsive forces between the atoms or molecules making up the medium. Such a potential for argon atoms or nitrogen molecules was first given a convenient analytic form by Lennard-Jones of Cambridge University, England, as a function of particle separation, r, and characteristic constants ϵ, σ which were properties of the molecular species:

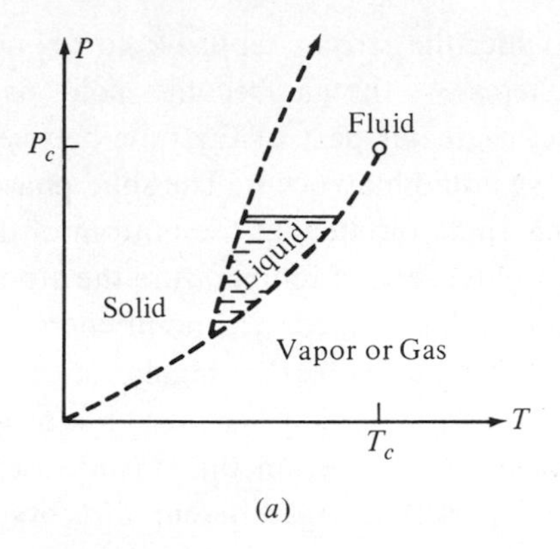

(a)

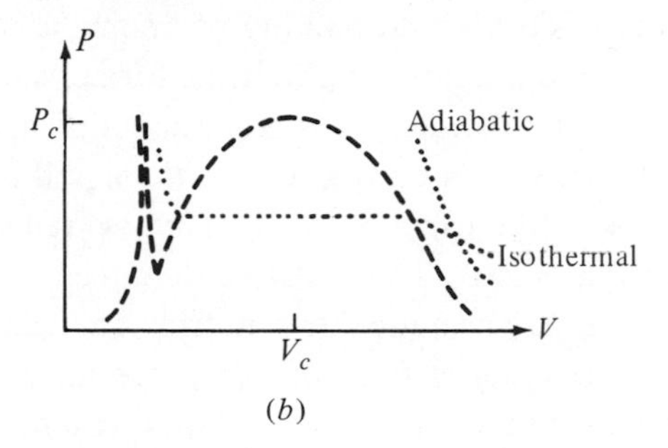

(b)

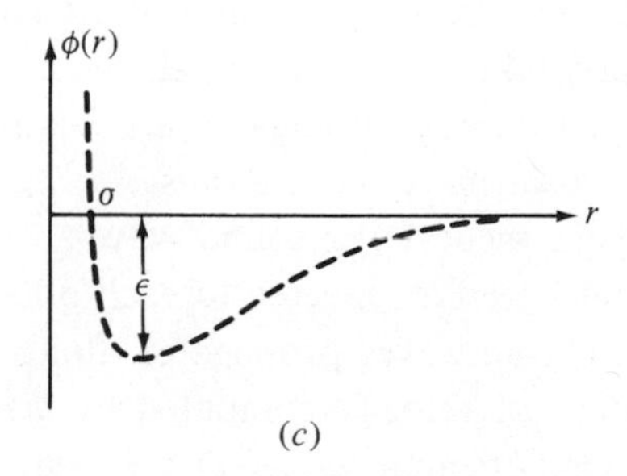

(c)

Figure 6.1a Plot of Pressure vs. Temperature for typical substance like argon. **b** Plot of Pressure vs. Volume for typical substance like argon. **c** Plot of Intermolecular Potential vs. Distance of Separation *r*.

$$\phi(r) = 4\epsilon\left[\left(\frac{\sigma}{r}\right)^{12} - \left(\frac{\sigma}{r}\right)^{6}\right] \tag{6.1}$$

The bottom of the potential well is not quite parabolic so the force law there is $F = -Kr + gr^2$ as in the anharmonic oscillator of Chapter 1. In the low pressure gas the particles are separated so far that only occasionally does the distance, r, become small enough for attractive forces to play a role.

When the molecules collide, the strong repulsive forces act as though the molecules were hard spheres. As the gas becomes cold, or more dense, the attractive forces may act a greater part of the time between collisions, and finally condensation to the liquid may occur. The solid phase is characterized by the molecule localized approximately at the bottom of the potential well, Fig. 6.1*c*. Even at the absolute zero of temperature the atoms oscillate about their equilibrium position with so-called zero-point energy.

In most simple solids the atoms form a regular lattice with equilibrium positions spaced in an orderly way. The macroscopic properties of such a solid may be understood to depend upon the crystal orientation; e.g., the compressibility may be different along different directions. The liquid in the region of Figs. 6.1*a* and 6.1*b* near the triple point has the ghost of lattice regularity in that there is a high probability of finding a near neighbor and a next nearest neighbor to an atom at about the former solid-phase distance between atoms. In a liquid the density is still quite high and a small change of volume makes the pressure rise rapidly; i.e., the liquid is not highly compressible in the region of the triple point. Finally, we remark that Fig. 6.1*a* shows a region called a "fluid". This must be described as a very dense gas whose equation of state may be quite different from an ideal gas. Generally, a fluid in this scientific meaning of the word does not have a meniscus separating the fluid and gas phases. However, a liquid taken to the region near the critical temperature and pressure becomes very compressible, has a much smaller density, and takes on the physical character of a fluid as described here. A fluid is characterized by having a given atom almost always in the potential field of two or more other atoms.

Newton tried to give a theory[2] for sound waves propagated in air by assuming that the molecules were connected to each other by a spring. The acoustic waves were like elastic waves propagated along the line of springs and mass points. The value of velocity computed by Newton was too low when compared to the experimental value, and it was not until 1822 that Laplace pointed out that the error lay in Newton's selection of the isothermal bulk modulus of air to describe the spring force constant rather than the adiabatic bulk modulus. The compression and expansion of the gas particles takes place so rapidly with respect to heat transport mechanisms that the process is nearly adiabatic, and regions of increased pressure (increased density) are regions of slightly higher temperature. In Fig. 6.1*b* we have shown a dotted line in the gas phase which has a steeper slope than the dashed isothermal shown there. The dotted line represents an adiabatic compression and

[2]Sir I. Newton, *Principia*, Book II, 1686.

expansion which is denoted by the well-known equilibrium thermodynamic equation

$$PV^{\gamma} = \text{constant} \tag{6.2}$$

From elementary texts on heat and thermodynamics we recognize that $\gamma = c_p/c_v$, the ratio of specific heat at constant pressure to the specific heat at constant volume. For a monatomic gas like argon, $\gamma = 5/3$ and for diatomic nitrogen, oxygen, etc., the value is 7/5. For a given quantity of gas at pressure P_0 and in a volume V_0 the constant in Eq. 6.2 can be written as $P_0 V_0^{\gamma}$ so that our equation becomes

$$PV^{\gamma} = P_0 V_0^{\gamma} \tag{6.3}$$

We may differentiate to obtain a very small pressure deviation

$$dP = -P_0 V_0^{\gamma} \gamma V^{-\gamma-1} dV \approx \gamma P_0 \frac{\Delta V}{V} \tag{6.4}$$

The quantity γP_0 can be called simply the bulk modulus of the gas. The excess pressure is proportional to the strain, which is like a simple Hooke's law of force between the atoms as used by Newton.

Let us now consider that the given quantity of gas occupies a short segment in a cylinder of constant cross-sectional area A as shown in Fig. 6.2. Imagine Fig. 6.2 is a long pipe with thick walls and of 16″ diameter with a loud speaker attached at one end and a small microphone (a hearing aid does well) mounted so that it may probe the standing wave pattern at any position along the tube.

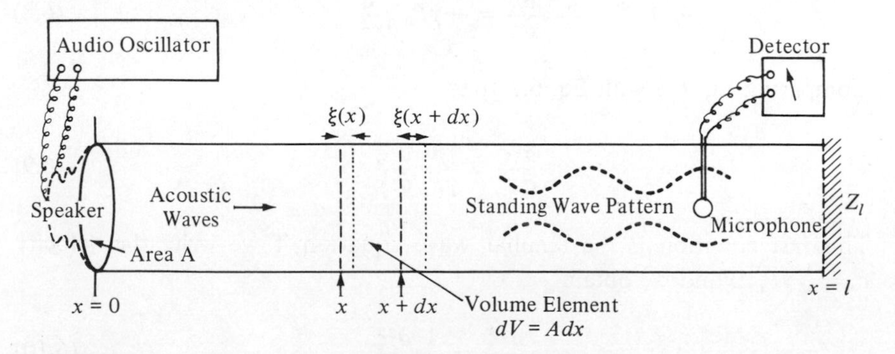

Figure 6.2 Acoustic Transmission Line showing Terminating Impedance and Microphone Detector to measure standing wave pattern.

Originally, the gas particles are found between x and $x + dx$ in a volume $A\,dx = V$. We must emphasize that there are some 10^{20} atoms in this small volume. After an adiabatic change, these same particles are found between $\xi(x)$ and $\xi(x + dx)$ where $\xi(x)$ is the particle displacement at the position x and $\xi(x + dx)$ is the somewhat different particle displacement at the position $x + dx$; i.e., the volume element has changed along the length of the tube. The volume change is

$$\Delta V = A[\xi(x + dx) - \xi(x)] = A\frac{\partial \xi}{\partial x}dx = V\frac{\partial \xi}{\partial x} \tag{6.5}$$

We may now substitute into Eq. 6.4 to give

$$dP = p = -\gamma P_0\frac{\partial \xi}{\partial x} \tag{6.6}$$

where the modulation of pressure is denoted by the small letter, p. We emphasize that Eq. 6.6 shows that p and ξ have maximum amplitude at different places. In a real physical system there is a time lag between the change in pressure and the adjusting movement of the atoms, and we have not taken this viscous damping effect into account.

Newton's law concerning force and acceleration requires that for a change of excess pressure along the length of the cylinder there must be an acceleration of the mass enclosed:

$$-\frac{\partial p}{\partial x} = \rho\frac{\partial^2 \xi}{\partial t^2} \tag{6.7}$$

If we differentiate Eq. 6.6 with respect to x, we have

$$\frac{\partial p}{\partial x} = -\gamma P_0\frac{\partial^2 \xi}{\partial x^2} \tag{6.8}$$

Comparing Eq. 6.8 with Eq. 6.7 gives

$$\frac{\partial^2 \xi}{\partial x^2} = \frac{\rho}{\gamma P_0}\frac{\partial^2 \xi}{\partial t^2} \tag{6.9}$$

This last equation is the familiar wave equation if we write the velocity $c = \sqrt{\gamma P_0/\rho}$ and we obtain

$$\frac{\partial^2 \xi}{\partial x^2} = \frac{1}{c^2}\frac{\partial^2 \xi}{\partial t^2} \tag{6.10}$$

An identical wave equation in the modulation of pressure, p, can be quickly derived from the above equations. We may differentiate Eq. 6.6 with respect to t and obtain

$$\frac{\partial^2 p}{\partial t^2} = -\gamma P_0 \frac{\partial}{\partial x}\left[\frac{\partial^2 \xi}{\partial t^2}\right] \tag{6.11}$$

But according to Eq. 6.7, we may substitute into this last equation to obtain

$$\frac{\partial^2 p}{\partial t^2} = -\gamma P_0 \frac{\partial}{\partial x}\left[-\frac{1}{\rho}\frac{\partial p}{\partial x}\right] = \frac{\gamma P_0}{\rho}\frac{\partial^2 p}{\partial x^2} \tag{6.12}$$

and this is the desired wave equation in pressure modulation about the steady pressure P_0 existing in the tube

$$\frac{\partial^2 p}{\partial x^2} = \frac{1}{c^2}\frac{\partial^2 p}{\partial t^2} \tag{6.13}$$

These linear and idealized wave equations for sound in a large tube are of course three-dimensional in that the cross section occupies the y and z coordinates and the wave propagates along the x-direction. Equations 6.10 and 6.13 are mathematically similar to the wave equation on a string. The amplitude of the string, $\psi(x)$, is similar to the amplitude of $\xi(x)$, but the maximum amplitude of p is not at the same position x and corresponds to $d\psi/dx$ or better to $d\xi/dx$. Another similarity is that the pressure wave theory has treated the ensemble of gas atoms as a continuum or continuous structure just as for the string and membrane. For frequencies below 10^8 cps the wavelength in the gas at STP is greater than 10^{-4} cm, and thus some 10^4 to 10^7 atoms may be within one wavelength. The kinetic theory of gases shows that at these ambient temperatures the *r. m. s.* velocity is of the order 10^4 cm/sec, and the gas may have a short mean free path. Local thermodynamic equilibrium is maintained by the frequent collisions between particles. For frequencies above 10^9 cps the question of a continuum theory remaining valid begins to arise and the question of the sound wave process remaining a constant entropy process must also be reexamined. According to Frisch [*Physics* **2**, 209 (1966)] and Berne, Boon, and Rice [*J. Chem. Phys.*, **47**, 2283 (1967)] under certain very general assumptions the linear response of a classical fluid becomes wholly nondissipative in the very high frequency limit; e.g., around frequencies of the order of the reciprocal collision time or some 10^{12} cps for liquid argon. Experimental evidence for this theoretical prediction has not yet been firmly established.

6.1 THE ACOUSTIC TRANSMISSION LINE

Pressure waves in a tube which is terminated at $x = L$ with a load impedance, Z_l, can be treated as a transmission line problem similar to the driven string treated earlier (see Fig. 6.2).

The pressure wave equation is:

$$\frac{\partial^2 p}{\partial x^2} = \frac{1}{c^2}\frac{\partial^2 p}{\partial t^2} \tag{6.13'}$$

The particle displacement wave equation is:

$$\frac{\partial^2 \xi}{\partial x^2} = \frac{1}{c^2}\frac{\partial^2 \xi}{\partial t^2} \tag{6.10'}$$

Let $\mathbf{k} = \omega/c$ be the wave vector, then the pressure wave moving to the right is:

$$p_+ = P_+ e^{i\mathbf{k}(x-ct)} \tag{6.14}$$

and the particle displacement for the outgoing wave is:

$$\xi_+ = A_+ e^{i\mathbf{k}(x-ct)} \tag{6.15}$$

The law that stress is proportional to strain was used in developing the wave equation and we repeat:

$$p = -\gamma P_0\left(\frac{\partial \xi}{\partial x}\right) = -\rho c \frac{\partial \xi}{\partial x} \tag{6.6'}$$

Therefore,

$$P_+ e^{i\mathbf{k}(x-ct)} = -\rho c^2 ik A_+ e^{i\mathbf{k}(x-ct)} \tag{6.16}$$

or

$$P_+ = -\rho c^2 ik A_+ = -i\omega\rho c A_+ \tag{6.17}$$

and we shall use the above equation to express A_+ in terms of P_+ and the other quantities.

Likewise, the waves moving to the left are:

$$p_- = P_- e^{-i\mathbf{k}(x+ct)} \tag{6.18}$$

$$\xi_- = A_- e^{-i\mathbf{k}(x+ct)} \tag{6.19}$$

and likewise,

$$A_- = \frac{P_-}{i\omega\rho c} \tag{6.20}$$

In the acoustic transmission tube, we have both waves moving to the right and reflected waves moving to the left:

$$p = P_+ e^{i\mathbf{k}(x-ct)} + P_- e^{-i\mathbf{k}(x-ct)} \tag{6.21}$$

Let the ratio of the amplitudes be written:

$$\frac{P_-}{P_+} = -e^{-2\pi\alpha_0 + i2\pi\beta_0} \qquad \text{at } x = 0 \tag{6.22}$$

Then we may rewrite the total pressure variations as:

$$p = P_+ e^{-\pi\alpha_0 + i\pi\beta_0}\{e^{\pi\alpha_0 - i\pi\beta_0 + i\mathbf{k}x} - e^{-\pi\alpha_0 + i\pi\beta_0 - i\mathbf{k}x}\}e^{-i\omega t} \tag{6.23}$$

or simply

$$p = 2P_+ e^{-\pi\alpha_0 + i\pi\beta_0} \sinh\{\pi\alpha_0 - i\pi\beta_0 + i\mathbf{k}x\}e^{-i\omega t} \tag{6.24}$$

Likewise, the total particle displacement is the sum of the wave to the right and the wave to the left:

$$\xi = A_+ e^{i\mathbf{k}(x-ct)} + A_- e^{-i\mathbf{k}(x+ct)} = \frac{P_+}{-i\omega\rho c} e^{i\mathbf{k}(x-ct)} + \frac{P_-}{i\omega\rho c} e^{-i\mathbf{k}(x+ct)} \tag{6.25}$$

The net particle velocity becomes:

$$\frac{\partial\xi}{\partial t} = \frac{P_+}{\rho c} e^{i\mathbf{k}(x-ct)} - \frac{P_-}{\rho c} e^{-i\mathbf{k}(x+ct)} \tag{6.26}$$

Substituting the value of P_- in terms of P_+ as above:

$$\frac{\partial\xi}{\partial t} = \frac{2P_+}{\rho c} e^{-\pi\alpha_0 + i\pi\beta_0} \cosh\{\pi\alpha_0 - i\pi\beta_0 + i\mathbf{k}x\}e^{-i\omega t} \tag{6.27}$$

The specific acoustic impedance at any point, x, in the tube is given by the ratio of pressure to particle velocity:

$$Z(x) = \frac{p}{\partial\xi/\partial t} = \rho c \tanh\{\pi\alpha_0 - i\pi\beta_0 + i\mathbf{k}x\} \tag{6.28}$$

By use of a microphone probe, we can measure the standing wave pressure amplitude at any position along the tube. In particular, the separation of $|p_{\max}|$ and $|p_{\min}|$ will be a quarter wave length, $\lambda/4$.

$$\text{The ratio} \qquad \frac{|p_{\min}|}{|p_{\max}|} = \tanh(\pi\alpha_0) = \frac{\sqrt{\cosh^2 \pi\alpha_0 - 1}}{\sqrt{\cosh^2 \pi\alpha_0}} \tag{6.29}$$

so by microphone measurement of the above pressure ratio we can calculate the quantity:

$$\alpha = \frac{1}{\pi} \tanh^{-1} \frac{|p_{\min}|}{|p_{\max}|} \tag{6.30}$$

The pressure minimum, $|p_{\min}|$, always occurs at a point where $\beta_0 - 2x/\lambda$ is an integer, $n = 0, 1$, etc.

The value at the load end of $\beta_0 - 2L/\lambda$ is called β_L. Just back from the load end is a *pressure minimum* to be found by the probe microphone at a distance back, $d = l - x$. Then at this minimum of pressure:

$$n = \beta_L + d/(\lambda/2) \tag{6.31}$$

We thus have directly measured β_L by selecting the proper integer value of $n = 0, 1$, etc., so that β_L will be a positive number between 0 and $\frac{1}{2}$. These values of α_0 and β_L are used in order to obtain the value of the impedance at the load:

$$Z(L) = \rho c \tanh [\pi(\alpha_0 - i\beta_L)] = \rho c\{\theta - i\chi\} \tag{6.32}$$

To calculate the mechanical impedance of the load we must multiply by the cross-sectional area of the tube. Thus, $Z_m = Z_L \cdot (\text{area})$. Even if the tube at $x = L$ is open to the external air, there is an abrupt change of the transmission line and this impedance mismatch causes some back reflection of the outgoing wave. The open organ-pipe was studied by Helmholtz (1860) and earlier by Lagrange, D. Bernouli, and Euler. Lord Rayleigh did experimental work on open tubes in 1877. A tube closed with a rigid plate gives $\alpha = 0$, $\beta_l = \frac{1}{2}$, the $Z(L) = \infty$ pure imaginary. This last terminating impedance is similar to a short-circuited electrical transmission line.

We may use the Smith Chart to calculate on this lossless transmission line or acoustic tube the impedance at the load end, $Z(L)$, as follows:

(*a*) Compute the ratio of maximum pressure to minimum pressure and mark this distance off on the scale of Standing Wave VOL Ratio. The pressures are experimentally given by the microphone probe and output meter.
(*b*) The distance, d, to the first *minimum* from the load toward the generator is measured and expressed as a certain fraction of a wavelength toward the generator, d/λ. The minimum of pressure is like the maximum for the string or electrical transmission line.
(*c*) The radial arm from the center of the chart (at 1.0 on the resistance component) is extended to a point on the circumference where the above fraction, d/λ, is located. The distance along this radial arm taken from part *a* above is the locus of the curves of resistive and reactive component of the terminal impedance. These values are read off to give R/Z_0 and jX/Z_0. Finally, $Z(L) = Z_0[R \pm jX]$ gives the acoustic impedance of the load.

(*d*) The Mechanical Impedance of the load is then, $Z_m = Z(L) \cdot$(area of tube).

6.2 LOSSES IN PLANE WAVE PROPAGATION: Shear Viscosity, Volume Viscosity, Heat Conduction

The plane waves in the rigid-walled pipe have been treated in the wave Eqs. 6.10 and 6.13 as idealized changes in the local variable. By considering a real gas whose physical properties are characterized by transport coefficients, an attenuation is obtained of the pressure wave amplitude as it moves along the transmission tube. Physically we can understand that the pressure wave has regions of slightly higher ambient temperature where compressed and regions adjacent of slightly lower temperature. The temperature gradient that exists allows energy to be transported from the higher to the lower temperature region. This energy taken out of the wave is lost to it and is dissipated to the medium. Also the wave motion is a compression and dilation together with some shear such that viscous friction effects cause a loss of energy from the wave to the medium.[3]

In the derivation of the ideal wave Eq. 6.13, the particles located in the volume element *Adx* of Fig. 6.2 were given a small displacement, small velocity, and were never far from thermodynamic equilibrium. If the gas in the tube were flowing parallel to the x-axis, there would be viscous friction so the velocity, $u(y)$, would increase in some way as one comes away from the tube wall. This gives rise to laminar flow and, according to Newton, the reactive force per unit area is proportional to the velocity gradient; the equation is

$$p_{yx} = -\eta_s \frac{\partial u}{\partial y} \tag{6.33}$$

The proportionality constant η_s depends on the nature of the gas, fluid, or liquid and is the familiar shear viscosity. Direct measurements of η_s may be obtained by the force measurements on a plate dragged through the medium. In general terms of a strain *rate* tensor, $\underline{\dot{\epsilon}}$, the stress tensor has a component of force per unit area due to shear motion given by

$$p_{ij} = -2\eta_s \dot{\epsilon}_{ij} \qquad \text{with} \left.\begin{matrix} i \\ j \end{matrix}\right\} = x, y, z \tag{6.34}$$

[3]See, for example, Arnold Sommerfeld, *Mechanics of Deformable Bodies*, page 82, Academic Press, New York (1964). Also see, L. D. Landau and E. M. Lifshitz, *Fluid Mechanics*, Addison-Wesley Publishing Co., Mass. (1959).

But with a compressible gas, liquid, or fluid we must add a term

$$p_{ij} = -\lambda\delta_{ij}\dot{\theta} \qquad \text{with } \delta_{ij} = \begin{matrix} 0 & \text{for } i \neq j \\ 1 & \text{for } i = j \end{matrix} \tag{6.35}$$

where $\dot{\theta} = \dot{\epsilon}_{xx} + \dot{\epsilon}_{yy} + \dot{\epsilon}_{zz}$ and λ is a second viscosity coefficient. The two terms above together give a force per unit area component

$$p_{ij} = -2\eta_s\dot{\epsilon}_{ij} - \lambda\delta_{ij}\dot{\theta} \tag{6.36}$$

In the case of uniform compression an isotropic frictional pressure of magnitude $(2\eta_s/3 + \lambda)\dot{\theta}$ arises and the number $(2\eta_s/3 + \lambda)$ is called the volume viscosity, η_v. Then $\lambda = (\eta_v - 2\eta_{s/3})$ and the complete Newtonian stress tensor (including the external pressure from the acoustic wave) is

$$\underline{\mathbf{P}} = -p\underline{\mathbf{1}} + (\eta_v - \tfrac{2}{3}\eta_s)\nabla\cdot u\underline{\mathbf{1}} + 2\eta_v\underline{\dot{\epsilon}} \tag{6.37}$$

where $\underline{\mathbf{1}}$ is the unit tensor. The tensor divergence $\nabla\cdot\underline{\mathbf{P}}$ produces a force vector which by Newton's equation of motion give

$$\nabla\cdot\underline{\mathbf{P}} = \rho\frac{d}{dt}\underline{\mathbf{u}} = \rho\frac{\partial u}{\partial t} + \rho\underline{\mathbf{u}}\cdot\nabla\mathbf{u} \tag{6.38}$$

which is the Navier-Stokes equation.

All of this tensor notation is made easy when we consider the plane wave in which products of small quantities, such as $\underline{\mathbf{u}}\cdot\nabla\rho$, $\underline{\mathbf{u}}\cdot\nabla\underline{\mathbf{u}}$, etc., can be neglected. The Navier-Stokes equation becomes

$$\rho\frac{\partial u}{\partial t} = -\frac{\partial p}{\partial x} + \left(\eta_v + \frac{4}{3}\eta_s\right)\frac{\partial^2 u}{\partial x^2} \tag{6.39}$$

or

$$\rho\frac{\partial^2\xi}{\partial t^2} = \gamma P_0\frac{\partial^2\xi}{\partial x^2} + \left(\eta_v + \frac{4}{3}\eta_s\right)\frac{\partial^3\xi}{\partial x^2\partial t} \tag{6.40}$$

where we have used Eq. 6.7 for the quantity $\partial p/\partial x$.

The last equation is just the wave equation altered with the viscosity term and is written again as

$$\frac{\partial^2\xi}{\partial t^2} = c^2\frac{\partial^2\xi}{\partial x^2} + \frac{(\eta_v + 4/3\eta_s)}{\rho}\frac{\partial^3\xi}{\partial x^2\partial t} \tag{6.41}$$

The damping of an oscillator in Chapter 1 was treated as a small perturbation and an attenuation determined, which was caused by the resistive term. We seek a similar solution to the wave equation and we make use of the experimental fact that the wave moves in the x-direction with particle displacement very nearly like Eq. 6.15, but modified to give

$$\xi = \xi_0 e^{-\alpha x} \cdot e^{ikx} \cdot e^{-i\omega t} \tag{6.42}$$

Differentiation and substitution into the new wave equation gives

$$-\omega^2 = c^2[\alpha^2 - 2ik\alpha - k^2] + \left(\frac{\eta_v + (4/3)\eta_s}{\rho}\right)(\alpha^2 - 2ik\alpha - k^2)(-i\omega) \tag{6.43}$$

Equating the imaginary parts equal to zero because the left side of the above equation is real, we have (assuming $\alpha \ll k$) the attenuation coefficient caused by friction

$$\alpha_\eta = \frac{\omega^2}{2\rho c^3}\left(\eta_v + \frac{4}{3}\eta_s\right) \tag{6.44}$$

From the real parts (assuming again that $\alpha \ll k$) we have $\omega^2 = c^2k^2$ − small terms. This dispersion relation does not contain the resonance conditions of the driven oscillator in Chapter 1, and the energy absorbed out of the wave does not become infinite at some characteristic resonance frequency ω_0. The medium could have a polyatomic molecule such as CO_2 whose quantum energy of rotation would easily pick up energy from the sound wave. L. E. Kinsler and A. R. Fry in *Fundamentals of Acoustics*, John Wiley and Sons (1950) give the following remark on page 262: "Many gases show the characteristic resonance attenuation due to molecular absorption. Pure CO_2 has a maximum molecular attenuation of 0.082 neper/cm at 20 kilocyles/sec which is 10^3 times classical attenuation." For CO_2 the characteristic temperature for rotation, $\theta_r = \hbar^2/2Ik = 0.8$ K. From the zeroth rotational energy level to the first level would require a quantum of energy, $h\nu$, such that $\nu = 1.67 \times 10^{10}$ cps. But at room temperature, which is large compared to 0.8 K, many upper rotational levels are occupied and the classical rotational energy of kT exists for the CO_2 molecule. Energy is taken out of the translational motion of the CO_2 during the sound wave compression, put into rotation, and the wave moves on before this energy is returned. Such a relaxation mechanism would add another term to the attenuation, and when added to the intrinsic volume viscosity may be given the general term "bulk viscosity." As Lord Kelvin once said, if we describe by theory a quantity such as volume viscosity, then we ought to measure it and give it a number. It is just from sound wave propagation studies that we can do this. But first we must describe another mechanism of attenuation caused by heat conduction so that we have the entire story.

The heat flowing out of a volume is caused by temperature gradients and we write

$$\int \rho \frac{\partial q}{\partial t} dV = K \int \text{grad } T \, dA = K \int \text{div grad } T \, dV \tag{6.45}$$

where q is the heat per gram and K is the thermal conductivity. By direct experiment in which a temperature gradient is established between parallel plates with a known heat flow, the value of K can be obtained for any substance at any absolute temperature and pressure. The change in heat content per gram of the gas is

$$dq = C_v \, dT + L_v \, dV = C_p \, dT + L_p \, dP \tag{6.46}$$

where C_v is the specific heat at constant volume, L_v is the latent heat per gram of volume increase,[4] C_p is the specific heat at constant pressure, and L_p the latent heat per gram with pressure increase. We shall consider a constant pressure process (nearly) so that Eq. 6.46 becomes

$$L_v = C_P \left(\frac{\partial T}{\partial V}\right)_P - C_v \left(\frac{\partial T}{\partial V}\right)_P = C_v(\gamma - 1)\left(\frac{\partial T}{\partial V}\right)_P \tag{6.47}$$

where $C_P/C_v = \gamma$.

The heat Eq. 6.45 becomes a "diffusion of temperature" equation

$$K\nabla^2 T = \rho c_v \frac{\partial T}{\partial t} + \rho C_v(\gamma - 1)\left(\frac{\partial T}{\partial \rho}\right)_P \frac{\partial \rho}{\partial t} \tag{6.48}$$

We write this last equation in a convenient form and hold the expression for future use:

$$\frac{\partial T}{\partial t} = -\frac{K}{\rho c_v} \nabla^2 T - (\gamma - 1)\left(\frac{\partial T}{\partial \rho}\right)_P \frac{\partial \rho}{\partial t} \tag{6.48$'$}$$

We account for the flow of particles out of a volume with the equation

$$\int \rho \dot{\xi} \, dA = -\frac{d}{dt} \int \rho \, dV \tag{6.49}$$

so that

$$\rho \frac{d}{dx} \dot{\xi} = -\frac{d\rho}{dt} \tag{6.49$'$}$$

and by differentiation we have

$$\frac{d}{dt}\left(\rho \frac{d}{dx} \dot{\xi}\right) = -\frac{d^2 \rho}{dt^2} \tag{6.50}$$

[4]See, for example, P. T. Landsberg, *Thermodynamics*, Interscience, New York (1961).

Newton's law of motion gives

$$\rho\ddot{\xi} = -\frac{dp}{dx} \tag{6.51}$$

and by differentiation we obtain

$$\frac{d}{dx}(\rho\ddot{\xi}) = -\frac{d^2p}{dx^2} \tag{6.52}$$

and, therefore, by comparing with Eq. 6.50 we have

$$\frac{d^2p}{dx^2} = \frac{d^2\rho}{dt^2} \tag{6.53}$$

We wish to use this last expression by putting the pressure into the variables ρ and T as follows:

$$dP = \left(\frac{\partial P}{\partial \rho}\right)_T d\rho + \left(\frac{\partial P}{\partial T}\right)_\rho dT \tag{6.54}$$

so that with differentiation we have

$$\frac{\partial^2 P}{\partial x^2} = \left(\frac{\partial P}{\partial \rho}\right)_T \frac{\partial^2 \rho}{\partial x^2} + \left(\frac{\partial P}{\partial T}\right)_\rho \frac{\partial^2 T}{\partial x^2} = \frac{\partial^2 \rho}{\partial t^2} \tag{6.55}$$

We may write this last equation as

$$\frac{\partial^2 \rho}{\partial t^2} = \left(\frac{\partial P}{\partial \rho}\right)_T \left[\frac{\partial^2 \rho}{\partial x^2} - \frac{1}{(\partial T/\partial \rho)_P} \frac{\partial^2 T}{\partial x^2}\right] \tag{6.55'}$$

where we have used an identity of partial derivatives.

The sound waves describing regions of slightly higher pressure, p, of density, ρ, and of temperature, T, all as a function of the variables x, ω, and t can be written

$$\begin{aligned} p &= p_0 e^{(ik-\alpha)x} \cdot e^{-i\omega t} \\ \rho &= \rho_0 e^{(ik-\alpha)x} \cdot e^{-i\omega t} \\ T &= T_0 e^{(ik-\alpha)x} \cdot e^{-i\omega t} \end{aligned} \tag{6.56}$$

The attenuation, α, will be determined by requiring that these last equations satisfy Eq. 6.48′ for heat conduction and Eq. 6.55′ for Newton's law of motion. Then by differentiation of Eq. 6.56 for the variable T, and also for the variable ρ, and substitution into Eq. 6.48′ we obtain

$$\left[(-i\omega) + \frac{K}{\rho c_v}(ik - \alpha)^2\right] T_0 = -(\gamma - 1)\left(\frac{\partial T}{\partial \rho}\right)_P (-i\omega)\rho_0 \tag{6.57}$$

Substitution into Eq. 6.55′ of the proper derivatives of T and ρ gives

$$\left[(-i\omega)^2 - \left(\frac{\partial P}{\partial \rho}\right)_T (ik - \alpha)^2\right]\rho_0 = -\left(\frac{\partial P}{\partial \rho}\right)_T \frac{1}{(\partial T/\partial \rho)_P}(ik - \alpha)^2 T_0 \tag{6.58}$$

Multiply Eq. 6.57 by Eq. 6.58 and obtain

$$\left[(-i\omega) + \frac{K}{\rho c_v}(ik - \alpha)^2\right]\left[(-i\omega)^2 - \left(\frac{\partial p}{\partial \rho}\right)_T (ik - \alpha)^2\right]$$
$$= [(\gamma - 1)(-i\omega)]\left[\left(\frac{\partial p}{\partial \rho}\right)_T (ik - \alpha)^2\right] \tag{6.59}$$

For $\alpha \ll k$ the attenuation coefficient due to heat conduction is

$$\alpha_c = \frac{(\gamma - 1)K}{c_p}\frac{\omega^2}{2\rho c^3} \tag{6.60}$$

Combining this last attenuation coefficient with that caused by viscosity Eq. 6.44, we have a total attenuation coefficient for the sound waves in the gas, fluid, or liquid given by*

$$\alpha = \alpha_\eta + \alpha_c = \frac{\omega^2}{2\rho c^3}\left[\eta_v + \frac{4}{3}\eta_s + \frac{(\gamma - 1)K}{c_p}\right] \tag{6.61}$$

By measuring the attenuation, α, of sound waves propagated as plane waves and by separate determination of all the other quantities which enter into the above expression, the value of intrinsic volume viscosity is calculated from Equation 6.61. It turns out that for simple liquids and fluids such as argon, nitrogen, oxygen, and methane, the value of η_v is nearly equal or slightly greater than that of η_s.

The theoretical studies on volume viscosity by Rice and Allnatt,[5] Rice and Gray,[6] and by Luks, Miller and Davis[7] give good agreement with the

*Landau and Lifshitz (see Ref. 3, page 298 in *Fluid Mechanics*) give another derivation involving the mean value of the energy dissipation and obtain Eq. 6.61. These authors call intrinsic volume viscosity "second viscosity," and in their Section 78 indicate that situations may exist to make second viscosity considerably greater than shear viscosity. These new situations are slow relaxation time mechanisms involving molecular internal degrees of freedom or any other energy exchange devices. We have called these "bulk viscosity."

[5]S. A. Rice and A. R. Allnatt, *J. Chem. Phys.* **34**, 2144 (1961).

[6]S. A. Rice and P. Gray, *The Statistical Mechanics of Simple Liquids*, John Wiley and Sons, New York (1965).

[7]See chapter on transport properties written by S. A. Rice, J. P. Boon, and H. T. Davis in *Simple Dense Fluids*, edited by H. L. Frisch and Z. W. Salsburg, Academic Press, New York (1968).

experimental results reported independently from four separate laboratories.[8] The rather lengthy discussion given in this book of attenuation by the medium of plane acoustic waves is justified by the knowledge gained of the transport properties of simple dense fluids.

PROBLEMS

1. The acoustic impedance tube has an area of cross section of 300 cm^2 and length 350 cm. The microphone probe indicates a pressure maximum of 12 dynes/cm^2 at points 347, 320, 293, etc. The pressure minimums are midway between these and have magnitude 5 dynes/cm^2. Use the Smith Chart to answer: What is the mechanical impedance at $x = 350$ cm, the load end?

2. Derive an expression for the wave equation in a conical horn of the type used by cheer-leaders at football games. Show that a solution for the space dependent part is

$$p(x) = \frac{1}{x}[P_+e^{ikx} + P_-e^{-ikx}]$$

[8]D. G. Naugle and C. F. Squire, *J. Chem. Phys.*, **42**, 3725 (1965) and **45**, 4669 (1966). D. Swyt, F. Havlice, and E. F. Carome, *J. Chem. Phys.*, **47**, 1199 (1967). W. M. Madigosky, *J. Chem. Phys.*, **46**, 444, (1967). A. E. Victor and R. T. Beyer, *J. Chem. Phys.*, **52**, 1573 (1970).

R. T. Beyer has given a thorough treatment of relaxation theory from the point of view of irreversible thermodynamics in his book: R. T. Beyer and S. V. Letcher, *Physical Ultrasonics*, Academic Press, New York (1969).

Chapter 7

SPHERICAL WAVES

The plane waves treated in the earlier chapters were useful for a description of many simple problems in the physics of particles. These plane waves may be expressed as spherical waves and the present chapter will develop this concept. In the treatment of forced motion of a membrane, the driving force was expressed in the characteristic functions of the membrane in order to understand the resulting motion. Likewise a beam of plane waves of sound can be expressed in the characteristic functions of a spherical coordinate system[1] if the physical problem at hand requires. For example, the scattering of a plane wave by a spherical object is treated this way. Beams of neutrons coming out of a reactor and made to scatter from target nuclei are similarly treated. For these reasons we express the acoustic wave equation (without attenuation) in spherical coordinates[2] and prepare the way for some mathematical tools used in the treatment of many problems in physics and engineering.

The wave equation for pressure in three dimensions is

$$\nabla^2 p = \frac{1}{c^2}\frac{\partial^2 p}{\partial t^2}$$

and may be written in the variables r, θ, and ϕ as

$$\left\{\frac{1}{r^2}\frac{\partial}{\partial r}\left(r^2\frac{\partial}{\partial r}\right) + \frac{1}{r^2\sin\theta}\frac{\partial}{\partial\theta}\left(\sin\theta\frac{\partial}{\partial\theta}\right) + \frac{1}{r^2\sin^2\theta}\frac{\partial^2}{\partial\phi^2}\right\}p = \frac{1}{c^2}\frac{\partial^2 p}{\partial t^2} \tag{7.1}$$

where we have transformed as follows:

[1] See, for example, Arnold Sommerfeld, *Partial Differenital Equations in Physics*, Academic Press, New York (1964).

[2] See, for example, E. Kreysig, *Advanced Engineering Mathematics*, 2nd edition, page 529, John Wiley and Sons, New York (1968).

$$x = r \sin \theta \cos \phi, \quad y = r \sin \theta \sin \phi, \quad z = r \cos \theta$$

With the background of the earlier chapters we know that the sound waves moving outward from the center came from some mechanical device there capable of setting the medium into oscillation at an angular frequency ω. If the medium is infinite in size there will be no return waves along the radius. The basic wave equation was derived quite independent of the type of sound source placed at the center and we must introduce our knowledge of this as a type of inner boundary condition. We must specify the source size, velocity amplitude, frequency, and displacement pattern. Texts (see Refs. 1 and 2) on the mathematics of the general solution of the wave equation proceed customarily with the separation of variables $p = R(r)\,\Theta(\theta)\,\Phi(\phi)\,T(t)$. We prefer to introduce the sound source at the very beginning and require the wave solution to meet the inner boundary conditions of the sound source.

7.1 SOUND SOURCE OF UNIFORMLY CONTRACTING AND EXPANDING SPHERE

We require the theoretical analysis to fit the physical system under study and we select the simple source of a puffing sphere as a starting point, Fig. 7.1. The pressure waves are symmetrical with respect to θ and ϕ so that the only space variation is with the variable r. The uniformly vibrating sphere of radius a is given a radial velocity at its surface

$$U = U_0 e^{-i\omega t} \tag{7.2}$$

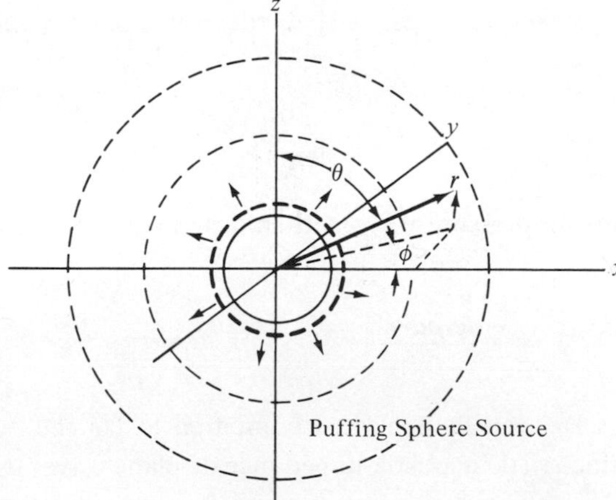

Figure 7.1 Puffing Sphere Sound Source.

and the gas or liquid medium at $r = a$ follows this velocity motion. The wave equation for distances $r > a$ becomes simply

$$\frac{1}{r^2}\frac{\partial}{\partial r}\left(r^2\frac{\partial p}{\partial r}\right) = \frac{1}{c^2}\frac{\partial^2 p}{\partial t^2} \tag{7.3}$$

The solution, which can be readily checked by differentiation, for the outward moving wave is

$$p = \frac{P_+}{r}e^{i\mathbf{k}r} \cdot e^{-i\omega t} \tag{7.4}$$

As the wave moves out along r, the amplitude gets smaller but the oscillations in r persist.

The particle velocity in the radial direction is

$$U_r = \frac{1}{i\omega\rho}\frac{\partial p}{\partial r} = \frac{ikP_+}{i\omega\rho r}e^{i(\mathbf{k}r-\omega t)} - \frac{P_+}{i\omega\rho r^2}e^{i(\mathbf{k}r-\omega t)} \tag{7.5}$$

or

$$U_r = \left(\frac{P_+e^{-i\omega t}}{i\omega\rho}\right)\left(\frac{e^{i\mathbf{k}r}}{r^2}\right)(ikr - 1) \tag{7.5$'$}$$

At $r = a$ where the particle velocity follows the puffing sphere we have

$$U_0 = \frac{P_+e^{i\mathbf{k}a}}{i\omega\rho a^2}[ika - 1] \tag{7.6}$$

For a small radius, a, compared to the acoustic wavelength, we have a very small fraction in the number $ka = \omega a/c = 2\pi a/\lambda$. Thus exp $(i\mathbf{k}a) \simeq 1$ and the term in the bracket above is $[-1]$. Under these simple approximations we have

$$U_0 = \frac{-P_+}{i\omega\rho a^2} \tag{7.7}$$

The outward moving pressure wave is then described in terms of the simple source sound as:

$$p = \frac{-i\omega\rho a^2 U_0}{r}e^{i(\mathbf{k}r-\omega t)} \qquad \text{for } r \geq a \tag{7.8}$$

At large r where $1/r^2$ can be neglected compared to $1/r$ the ratio $p/u \simeq \rho c$ which is the characteristic acoustic impedance of plane waves used in the last chapter. The spherical wave at great distance from any general source may, as an approximation, be taken as Eq. 7.8. This approach is used in optical waves from a point source and in other systems. The amplitude of pressure and amplitude of particle velocity falls off as $1/r$ at these large distances so that the signal intensity, $\overline{p \cdot u}$, averaged over one cycle, falls off as $1/r^2$. For an experiment covering, say, 20 cm of length at a distance of 100 meters from the source, the pure plane wave approximation without a $1/r$ factor is justified at the experimental site.

7.2 SOUND SOURCE OF RIGID SPHERE VIBRATING ALONG z-AXIS—DIPOLE SOURCE: Legendre Polynomials, Bessel Functions

Figure 7.2 shows a small rigid sphere which vibrates along the z-axis so that the velocity at its surface is

$$U = U_0 \cos\theta \, e^{-i\omega t} \tag{7.9}$$

The variable, $\cos\theta$ runs from 1 to -1; the radius r runs from a to ∞. It is clear that the pressure waves generated by this dipole source are symmetric in the azimuth angle ϕ so that the wave equation suitable for this situation is

$$\frac{1}{r^2}\frac{\partial}{\partial r}\left(r^2\frac{\partial p}{\partial r}\right) + \frac{1}{r^2 \sin\theta}\frac{\partial}{\partial \theta}\left(\sin\theta\frac{\partial p}{\partial \theta}\right) = \frac{1}{c^2}\frac{\partial^2 p}{\partial t^2} \tag{7.10}$$

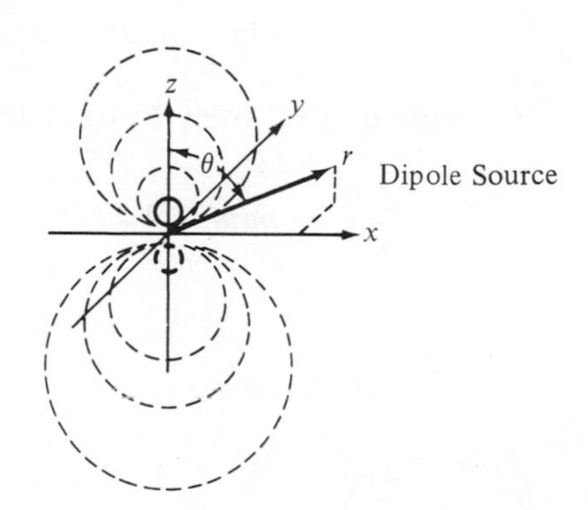

Figure 7.2 Dipole Sound Source.

The solution we try involves a separation of the variables

$$p = R(r)\,\Theta(\theta)e^{-i\omega t} \tag{7.11}$$

Differentiation and substitution back into the wave equation is just the technique so often used and gives

$$\frac{1}{R}\frac{\partial}{\partial r}\left(r^2\frac{\partial R}{\partial r}\right) + \frac{\omega^2}{c^2}r^2 = -\frac{1}{\Theta}\frac{1}{\sin\theta}\frac{\partial}{\partial\theta}\left(\sin\theta\frac{\partial\Theta}{\partial\theta}\right) \tag{7.12}$$

The left side is a function of r only so must be equal to a constant C. The right side is also equal to the same constant and is

$$\frac{1}{\sin\theta}\frac{\partial}{\partial\theta}\left(\sin\theta\frac{\partial\Theta}{\partial\theta}\right) + \Theta C = 0 \tag{7.13}$$

This equation may be reduced[3] by the substitution $\cos\theta = \mu$, which is the variable used in the present dipole source:

$$(1-\mu^2)\frac{d^2\Theta}{d\mu^2} - 2\mu\frac{d\Theta}{d\mu} + C\Theta = 0 \tag{7.13'}$$

The substitution made used of the following:

$$-\sin\theta\, d\theta = d\mu, \quad \sin\theta\frac{d}{d\theta} = \sin^2\theta, \quad \frac{d}{\sin\theta\, d\theta} = -(1-\mu^2)\frac{d}{d\mu}$$

Equation 7.13′ is known as Legendre's equation and to solve it we use the series method as in Chapter 5, Eq. 5.30, where the resulting series defined a Bessel function of the first kind of order m. In the present situation we shall obtain the Legendre polynomials.[4]

$$\Theta = a_0\mu^m + a_1\mu^{m+1} + a_2\mu^{m+2} + \cdots \tag{7.14}$$

Differentiation in order to form the terms in Eq. 7.13′ gives

$$\begin{aligned}
\frac{d^2\Theta}{d\mu^2} &= m(m-1)a_0\mu^{m-2} + (m+1)ma_1\mu^{m-1} \\
&\qquad + (m+2)(m+1)a_2\mu^m + \cdots \\
-\mu^2\frac{d^2\Theta}{d\mu^2} &= \qquad -m(m-1)a_0\mu^m - \cdots \\
-2\mu\frac{d\Theta}{d\mu} &= \qquad -2ma_0\mu^m - \cdots \\
C\Theta &= \qquad Ca_0\mu^m + \cdots
\end{aligned}$$

The sum of these terms must be zero and hence we must have

$$\begin{aligned}
&m(m-1)a_0 = 0, \quad (m+1)ma_1 = 0, \\
&\qquad (m+2)(m+1)a_2 - [m(m-1) - 2m + C]a_0 = 0
\end{aligned}$$

We see that if a_0 is arbitrary that m must be either 0 or 1. If we take $m = 0$ then the next equation involving the series coefficents shows that a_1 is arbitrary and the third equation shows

$$a_2 = -\frac{C}{2}a_0$$

[3]See, for example, H. Bateman, *Partial Differential Equations of Mathematical Physics*, page 351, Dover Publications, New York (1944).

[4]See, for example, F. S. Wood, *Advanced Calculus*, Ginn and Co., New York (1932).

The coefficients a_3, a_4, etc., can also be determined from the higher terms. We can construct the result that

$$\Theta = a_0\left[1 - \frac{C}{2!}\mu^2 - \frac{C(6-C)}{4!}\mu^4 - \frac{C(6-C)(20-C)}{6!}\mu^6 - \cdots\right]$$
$$+ a_1\left[\mu + \frac{(2-C)}{3!}\mu^3 + \frac{(2-C)(12-C)}{5!}\mu^5 + \cdots\right] \tag{7.15}$$

The only way we can aviod divergence of these two series at $\mu = \pm 1$, which is a value the variable may take, is to restrict the value of C to

$$C = n(n+1) \qquad n = 0, 1, 2, 3, \text{etc.}$$

In this way a value of n makes one series finite and the other series is made to vanish by requiring its coefficient, a, to be zero. The polynomial for a given value of n is then further specified by making the remaining coefficient, a, such that

$$\Theta_n(\mu) = 1 \qquad \text{at } \mu = 1$$

These special polynomials, Legendre polynomials, are customarily given a special symbol $P_n(\mu)$ rather than $\Theta_n(\mu)$.

Values for a few of these are:

$$P_0(\cos\theta) = 1$$
$$P_1(\cos\theta) = \cos\theta$$
$$P_2(\cos\theta) = \tfrac{1}{4}(3\cos 2\theta + 1)$$

and each satisfies a Legendre differential equation in which the value of n is denoted by the subscript. Evidently, $P_n(\cos\theta)$ emerges as a characteristic function which exactly fits the nth mode of motion of the vibrating source with which we have been dealing. $P_0(\cos\theta) = 1$ is the simple puffing sphere which is quite symmetric in the polar angle θ, and $P_1(\cos\theta) = \cos\theta$ describes the dipole source under present investigation. P_2 $(\cos\theta)$ is the correct characteristic function for a "quadrupole" source, and we do not suggest some mechanical way to produce this source.

We must turn our attention to the radial part of the solution, $R(r)$, in order to match it to the value n chosen for the Legendre function. Equation 7.12 has a radial equation which is

$$\frac{1}{R}\frac{\partial}{\partial R}\left(r^2\frac{\partial R}{\partial r}\right) + \frac{\omega^2}{c^2}r^2 = n(n+1) \tag{7.16}$$

or $$\frac{d^2R}{dr^2} + \frac{2}{r}\frac{dR}{dr} + \left(k^2 - \frac{n(n+1)}{r^2}\right)R = 0 \qquad \text{with } k = \omega/c \tag{7.17}$$

For the dipole $n = 1$ and this becomes

$$\frac{d^2R}{dr^2} + \frac{2}{r}\frac{dR}{dr} + \left(k^2 - \frac{2}{r^2}\right)R = 0 \tag{7.18}$$

The variable combination kr goes together (as does kx in plane waves) and we define the symbol $z = kr$ to change Eq. 7.18 into

$$\frac{d^2R}{dz^2} + \frac{2}{z}\frac{dR}{dz} + \left(1 - \frac{2}{z^2}\right)R = 0 \tag{7.18$'$}$$

We take the series solution method again and write

$$R = a_0 z^m + a_1 z^{m+1} + a_2 z^{m+2} + a_3 z^{m+3} + a_4 z^{m+4} + \cdots \tag{7.19}$$

$$\frac{d^2R}{dz^2} = m(m-1)a_0 z^{m-2} + (m+1)ma_1 z^{m-1} + (m+2)(m+1)a_2 z^m + (m+3)(m+2)a_3 z^{m+1} + \cdots$$

$$\frac{2}{z}\frac{dR}{dz} = 2ma_0 z^{m-2} + 2(m+1)a_1 z^{m-1} + 2(m+2)a_2 z^m + 2(m+2)a_3 z^{m+1} + \cdots$$

$$-\frac{2}{z^2}R = -2a_0 z^{m-2} - 2a_1 z^{m-1} - 2a_2 z^m - 2a_3 z^{m+1} - \cdots$$

From the sum of the coefficients of the terms in z^{m-2} we have

$$[m(m+1) + 2m - 2]a_0 = 0$$

If a_0 is to be arbitrary, then the terms in the bracket must be zero and the resulting quadratic equation gives values of $m = 1$ or $m = -2$. From the sum of the coefficients of the terms in z^{m-1} we have $[(m+1)m + 2(m+1) - 2]a_1 = 0$, and a_1 must be identically zero for the values of both $m = 1$ and for $m = -2$. Continuing, we have

$$m = 1, \quad a_2 = -a_0/10, \quad a_4 = a_0/260, \quad a_3 = a_5 = 0$$

$$m = -2, \quad a_2 = a_0/2, \quad a_4 = -a_0/4, \quad a_3 = a_5 = 0$$

For $m = 1$ we have the series for a particular solution

$$R = a_0\left[z - \frac{z^3}{10} + \frac{z^5}{260} + \cdots\right] = a_0\frac{z^{3/2}}{z^{1/2}}\left[1 - \frac{z^2}{10} + \frac{z^4}{260} - \cdots\right] \tag{7.20}$$

We recognize that Eq. 7.20 can be expressed by the Bessel function whose general expression is

$$J_n = \frac{z^n}{2^n\Gamma(n+1)}\left[1 - \frac{z^2}{2(2n+2)} + \frac{z^4}{2\cdot 4(2n+2)(2n+4)} - \cdots\right]$$

Taking the value of $n = 1 + 1/2$, we have (with $\Gamma(5/2) = 3\sqrt{\pi}/4$) the solution

$$R = \frac{A}{\sqrt{z}} J_{(1+1/2)}(z) \tag{7.21}$$

For the condition $m = -2$, we have the series

$$R = a_0\left[z^{-2} + \frac{1}{2} - \frac{z^2}{4} + \cdots\right] \tag{7.22}$$

Obviously for large $z = kr$ this series is small compared to that denoted by $m = 1$, but for small z this series dominates. Thus for $kr \longrightarrow 0$ (but r itself not less than a) we have

$$p \simeq AP_1(\cos\theta)\frac{1}{z^2} \tag{7.23}$$

We shall need the differentiation with respect to r and write

$$\frac{\partial p}{\partial r} \simeq -2AP_1(\cos\theta)\frac{1}{z^3} \tag{7.24}$$

The dominant solution for the pressure wave given off by the dipole at reasonably large values of kr is

$$p = A\cos\theta\frac{J_{1+1/2}(\mathbf{k}r)}{\sqrt{\mathbf{k}r}}e^{-i\omega t} \tag{7.25}$$

or

$$p = AP_1(\cos\theta)\frac{J_{1+1/2}(\mathbf{k}r)}{\sqrt{\mathbf{k}r}}e^{-i\omega t} \tag{7.25$'$}$$

The amplitude of the wave decreases as the wave moves out along r because of the factor $1/\sqrt{kr}$ and because of the amplitude decrease of the Bessel function. The oscillations of the pressure as r increases are not periodic as in Eq. 7.4, but at great distance the Bessel function approaches $\cos(kr - \delta)/\sqrt{kr}$ where δ is a phase shift to make the cosine function fit the Bessel function. Thus at great distances the pressure falls off as $1/r$ and the oscillations are periodic in kr just as in Eq. 7.8. More precisely, we write the pressure at great distance from the source as

$$p = -A\frac{e^{i\mathbf{k}r}}{\mathbf{k}r}\cos\theta e^{-i\omega t} \tag{7.26}$$

where the minus sign accounts for the phase shift 180° and we have used the periodic function exp($i\mathbf{k}r$).

The radial particle velocity at small distances (or small kr) is obtained with the use of Eq. 7.24

$$u = \dot{\xi} = \frac{1}{i\omega\rho}\frac{\partial p}{\partial r} = \left(\frac{2A}{\rho c}\right)\left(\frac{1}{kr}\right)^3 \cos\theta e^{-i\omega t} \tag{7.27}$$

If this value is taken at $r = a$, then we equate the source velocity to the particle velocity to obtain

$$U_0 \simeq \left(\frac{2A}{\rho c}\right)\left(\frac{1}{ka}\right)^3 \tag{7.28}$$

or

$$A \simeq \frac{4\pi^3 \rho \nu^3 a^3 U_0}{c^2} \tag{7.29}$$

The sound intensity $\overline{p \cdot u}$ becomes at large distance

$$I \simeq \frac{2\pi^4 \rho \nu^4 a^6 U_0^2}{c^3} \frac{\cos^2\theta}{r^2} \tag{7.30}$$

This result for a dipole source of pressure waves is similar to a dipole source of electromagnetic radiation, and it is well to remember that the intensity goes as the fourth power of frequency, as the dipole strength squared (U_0^2), and as $1/r^2$. The approximations are valid when the radius, a, of the vibrating sphere is much less than a wavelength of sound.

The pressure for the nth mode of motion of the source can be written

$$p = A_n P_n(\cos\theta)\frac{J_{n+1/2}(\mathbf{k}r)}{\sqrt{\mathbf{k}r}} \tag{7.31}$$

It is possible to think of some radiating source at angular frequency ω, whose space configuration is made up partly with $n = 0$ and partly with $n = 1$, $n = 2$, etc. A series with the amplitude coefficients adjusted to fit the source would be required. In a manner of speaking, G. Stokes did this in 1868 and his theory is given in Lord Rayleigh's *The Theory of Sound.*[5] Stokes used an expression

$$R = \left(\frac{e^{i\mathbf{k}r}}{\mathbf{k}r}\right) A_n(\mathbf{k}r)$$

[5]Volume II, Chapter XVII, Dover Publications, New York (1945).

to solve Eq. 7.18′ and got an expression for the function $A(\mathbf{k}r)$ in a polynomial in powers of $(1/kr)$. For a puffing sphere, $A(\mathbf{k}r)$ is a constant.

It is interesting that a plane wave can be expanded in these functions

$$e^{i\mathbf{k}r\cos\theta} = \sum_{n=0}^{\infty} A_n \frac{J_{n+1/2}(\mathbf{k}r)}{\sqrt{\mathbf{k}r}} P_n(\cos\theta) \tag{7.32}$$

Arnold Sommerfeld in *Partial Differential Equation in Physics*, page 144, shows that the coefficients A_n can be calculated from the orthogonality of the P_n. Sommerfeld, some of whose students became Nobel Laureates and presidents of universities, considered these mathematical tools of great importance in physics and engineering.

P. M. Morse[6] discovered that Eq. 7.18′ can have two solutions:

$$j_1(z) = \frac{\sin z}{z^2} - \frac{\cos z}{z} \qquad \text{called a spherical Bessel function}$$

$$n_1(z) = -\frac{\sin z}{z} - \frac{\cos z}{z^2} \qquad \text{called a spherical Neumann function}$$

The spherical Bessel function is defined in terms of the ordinary Bessel function

$$j_1(z) = \sqrt{(\pi/2z)}J_{1+1/2}(z)$$

The dipole produces a pressure wave in these functions

$$p = AP_1(\cos\theta)[j_1(kr) + in_1(kr)]e^{-i\omega t}$$

7.3 SOUND SOURCE OF VIBRATING PISTON:

Huygens Principle, Fraunhofer Diffraction

In this section we deal with sound waves produced by a source that can be generally treated as a vibrating piston, but which may be an electromagnetic device used for producing high frequency or very high frequency (ultrasonic) waves. The great French physicist, Paul Langevin (1872-1946) worked on ultrasonic echo-ranging for detection of enemy submarines during World War I, and his device received a British patent in 1921. Langevin used a quartz crystal whose property is such that it expands and contracts as alternating voltage is applied to its surfaces when the quartz is the dielectric medium between the plates of a condensor. Hence, as high-frequency voltage is

[6] P. M. Morse, *Vibration and Sound*, McGraw-Hill Book Co., New York, (1948).

applied for several microseconds up to a millisecond, ultrasonic waves are emitted by the vibrating surface. The crystal can also produce an alternating voltage across its opposite surfaces when excited by mechanical vibrations of the return pressure waves or echo. These electrical signals of the return echo can be timed with respect to the initial pulse and, from the known speed of sound in sea water, measurement made on the range and direction of the underwater object. The technique of radar ranging is identical to sonar ranging in many ways. This section will deal with the directional or beam character of the high-frequency sound waves emitted by the piston-like action of the quartz transducer. Incidentally, the Germans put Langevin in prison during World War II because he was regarded as dangerous to their conquest efforts. Piston-driving action from magnetostriction devices and many other piezoelectric crystals, such as Rochelle salt and ceramic barium titanate, have been used in submarine detection work. They are housed in protecting sound domes as shown in the photograph of Chapter 6. In Chapter 6 we discussed the attenuation of sound waves by the medium. Experimental knowledge of this comes also from ultrasonic pulse-echo methods where the intensity of the return echo is measured in decibels. These practical and scientifically important properties of the vibrating-piston source require our attention.[7]

Figure 7.3*a* shows a schematic of the quartz crystal transducer mounted in a typical liquid test cell with a movable target reflector. Figure 7.3*b* is an enlarged view of the front surface of the crystal and the coordinate system used to describe the radiating pressure waves. The pressure, dp, at a point P located at some distance r from the vibrating piston is given by Eq. 7.8 and is caused by the motion of the surface element dA. We write

$$dp = \frac{-i\omega\rho U_0}{r'} e^{i(\mathbf{k}r' - \omega t)} \tag{7.33}$$

The piston has velocity $U = U_0 \exp(-i\omega t)$ uniformly across the face. The distance from the element of area dA to the point in space is r' which is only slightly different than the distance r from the center of the piston to the point P. That small difference does not matter so much as the fact that the pressure wave from another element of area dA' located at some other region of the piston face can arrive at point P with the same amplitude but just 180° out of phase or just one-half a wavelength behind. It is just this destructive (and constructive) interference effect that causes the narrow beam shape of the

[7] An advanced text on these topics is by R. T. Beyer and S. V. Letcher, *Physical Ultrasonics*, Academic Press, New York (1969).

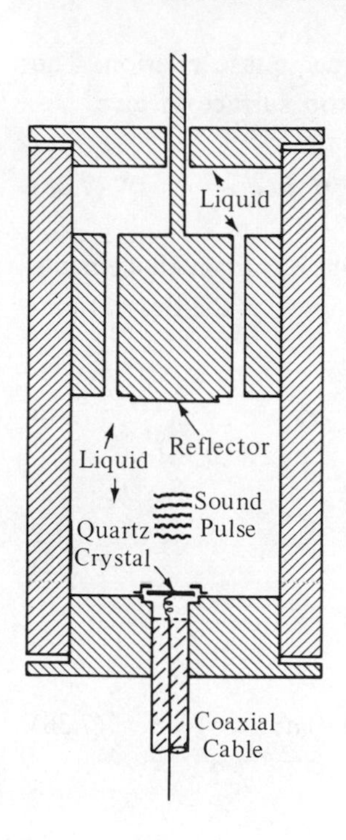

Figure 7.3a Piston Sound Source made by Quartz Crystal shown in ultrasonic experimental cell with reflector for echo signal.

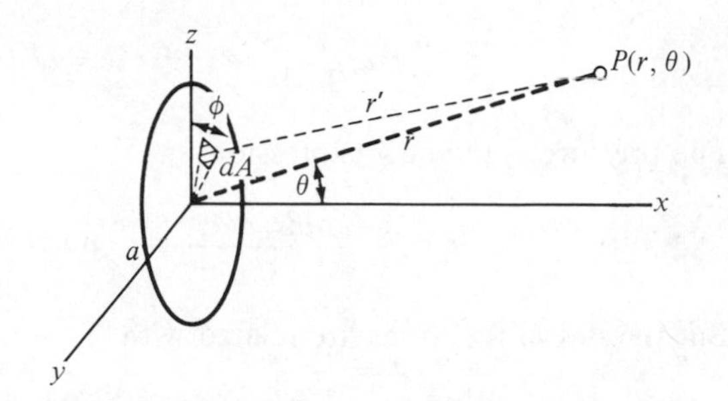

Figure 7.3b Piston Sound Source showing coordinates used in theory.

sound radiated by the piston at high frequencies. Let the radius of the piston be denoted by σ. Then the geometry of the system is such that

$$r'^2 - r^2 = \sigma^2 - 2r\sigma \cos \phi \sin \theta \tag{7.34}$$

Factoring the left side gives the approximation

$$(r' - r)(r' + r) \simeq (r' - r)\, 2r$$

because r' is nearly equal to r. Furthermore, $\sigma^2/2r$ is small compared to $\sigma \cos \phi \sin \theta$. Thus, neglecting small terms, we write Eq. 7.34 as

$$r' = r - \sigma \cos \phi \sin \theta \tag{7.34'}$$

We may put this value of r' into Eq. 7.33 and in the denominator of that equation we can simply use r for r', but in the exponent the complete expression of Eq. 7.34′ must be retained to give the proper phase relation. The total pressure at point P is integrated over the piston surface to give

$$p = \frac{-i\rho\omega U_0 e^{i(\mathbf{k}r-\omega t)}}{r} \int_0^a \int_0^{2\pi} \sigma e^{-ik\sigma \sin\theta \cos\phi}\, d\sigma\, d\phi \tag{7.35}$$

The process is a Huygens principle of adding contributions from all elements of radiation. The integral (see Ref. 1 and 2 of this chapter) is the Bessel function:

$$\frac{1}{2\pi i^m} \int_0^{2\pi} e^{iz\cos\Omega} \cos(m\Omega)\, d\Omega = J_m(z) \tag{7.36}$$

In the case $m = 0$ this is

$$\frac{1}{2\pi} \int_0^{2\pi} e^{iz\cos\Omega}\, d\Omega = J_0(z) \tag{7.37}$$

The pressure is the new expression

$$p = \frac{-i\omega\rho U_0 e^{i(\mathbf{k}r-\omega t)}}{r} \int_0^a \sigma J_0(k\sigma \sin\theta)\, d\sigma \tag{7.38}$$

But the Bessel functions are related with

$$\int zJ_0(z)\, dz = zJ_1(z) \tag{7.39}$$

We therefore have the final expression for the pressure whose real part can be determined:

$$p = \frac{-i\rho\omega U_0 a^2}{2r} \left[\frac{2J_1(ka\sin\theta)}{ka\sin\theta}\right] e^{i(\mathbf{k}r-\omega t)} \tag{7.40}$$

The first zero of the quantity in the bracket is at $ka \sin\theta = 1.2\pi$ so that

$$\sin\theta = \frac{1.2\pi}{ka} = \frac{0.61\lambda}{a} \tag{7.41}$$

and there is very little radiation outside this angle because of the rapid damping of the amplitude of the Bessel function expression in Eq. 7.40.

The order of magnitude for the angle θ can be given for a liquid with sound velocity 800 meters/sec and at frequency of 10 megacycles/sec from a quartz crystal of radius $a = 1$ cm:

$$\sin\theta = 0.61\,(8 \times 10^4/10^7) = 4.8 \times 10^{-3}$$

The lobe containing most of the sound energy has an angular width in this case of about 0.6°.

For these reasons the ultrasonic energy is in a very well defined beam and is nearly a plane wave after traveling several hundred wavelengths from the piston source. If the wave is then reflected from a flat surface as in Fig. 7.3*a*, the sound pulse is returned to the transducer with loss due to the

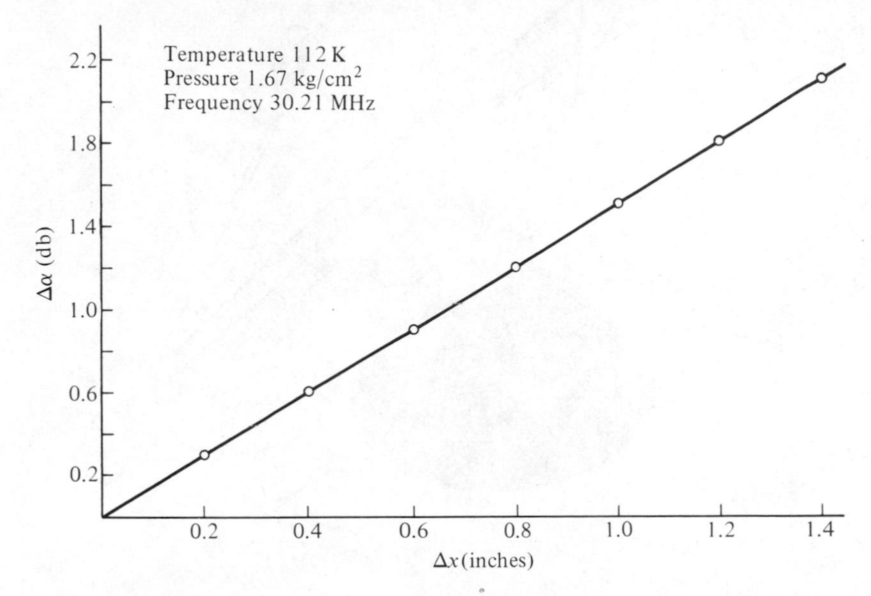

Figure 7.4 Ultrasonic Attenuation vs. Distance (round trip) to Reflector for liquid methane. The sound pressure level is the decibel (db), and changes in pressure level Δ(db) are measured such that $\Delta L_p = 20 \log (p_1/p_2)$ in units of Δ(db). The straight line plot proves experimentally that Eq. 6.56 is valid; i.e., $P \approx P_0 e^{-\alpha x}$. Thus $\alpha = (1/x)\ ln\ (P_0/P)$ and ΔL_p(db/cm) = 8.686α (Neper/cm).

liquid medium. Figure 7.4 shows the loss in decibels plotted against the round-trip distance from the transducer to the reflector.[8] In this example the medium is liquid methane. In Chapter 6 we saw that the attenuation increased with the square of the frequency. But at low frequencies the beam width is large, so directivity is lost. Even with a transducer having an effective width of 3 meters as we might imagine from the drawing, Fig. 7.5, inspired by the photograph in Chapter 6, the need for a well defined beam for direction-finding purposes would require frequencies of the order 7×10^4 cycles/sec. The directional effect for the high audio frequencies of 10^4 cycles/sec can be troublesome in loud speakers because people off to the side of the main lobe are not given all the tonal qualities of the music. An array of speakers can remove this difficulty.

[8]J. R. Singer, *J. Chem. Phys.*, **51**, 4729 (1969).

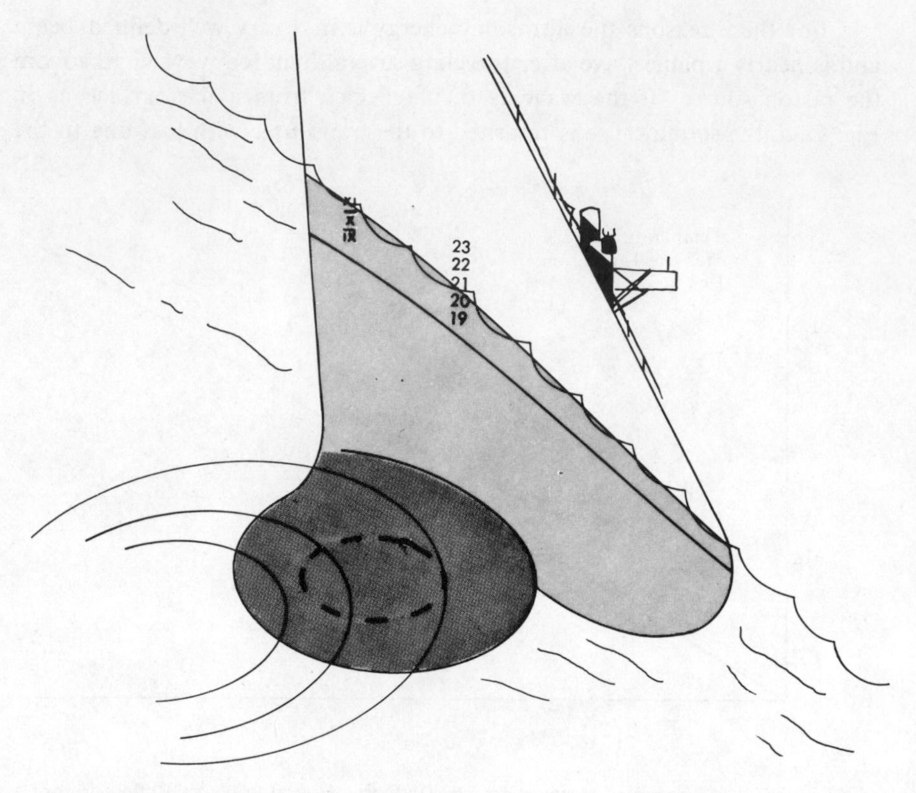

Figure 7.5 Drawing of Sonar Beam inspired by photo in Chapter 6.

Finally, we remark that the theory for Fraunhofer diffraction of light by a circular aperture is mathematically identical to the above treatment for the radiation by a piston. Huygens' principle applies to the incident light wave coming through the aperture. Behind the aperture we get a very intense central beam with space spread $\sin\theta = 0.61\lambda/a$ as in Eq. 7.41. The scalar wave (like Eq. 7.40) can be used even though there must be a vector wave **E** in electromagnetic radiation. In point of fact we would be hard put to measure anything but a scalar quantity from a distant point source of light caused by a small flame. Even if we could measure at an instant the magnitude and direction of the electric vector **E** at the point P, we would have no idea about **E** in the next instant of time as to phase and magnitude. The square of the scalar wave amplitude gives the intensity at point P so we have (with I_0 the intensity at the center of the diffraction pattern):

$$I(P) = I_0\left[\frac{2J_1(ka\sin\theta)}{ka\sin\theta}\right]^2$$

Chapter **8**

SPHERICAL WAVES IN QUANTUM SYSTEMS

In this chapter we apply the mathematical apparatus developed for spherical waves in Chapter 7 to a quantum system. Our knowledge of spherical harmonics will be further developed in doing this. It has been our procedure to describe the physical system and tell what can be observed or measured. Then we have followed through with the theoretical attack which has as its goal an explanation capable of predicting the results of any further observation or measurement. The background assumed here on the part

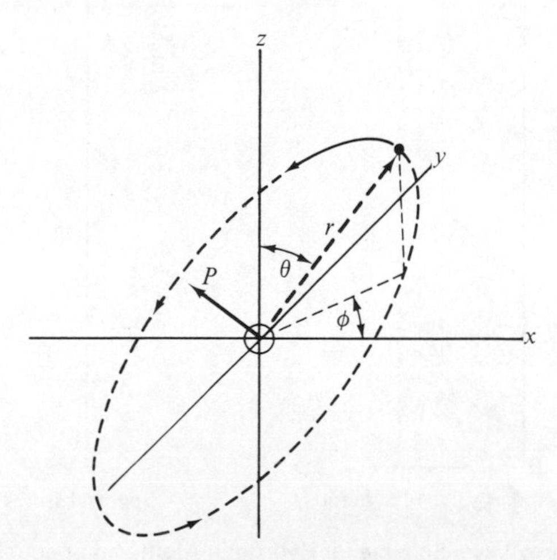

Figure 8.1 Simple Planetary Model of the Hydrogen Atom. The volume per mol of liquid H_2 is 26.4 cm^3 and there are 6×10^{23} molecules per mol so the volume of each molecule is about 4.4×10^{-23} cm^3. The radius of the H_2 is therefore about 1×10^{-8} cm, and the radius of the hydrogen atom is about $\frac{1}{2} \times 10^{-8}$ cm.

of the reader is that he knows that matter is made up of atoms of the chemical elements, that the atoms are electrically neutral, and that they contain negatively charged electrons which form a cloud about a positively charged mass located in a small region at the center of the atom (E. Rutherford, *Phil. Mag.*, **21**, 669, 1911). Figure 8.1 gives the planetary model of the hydrogen atom which early scientists of this century used to explain the experimental results. Students at the level of the present text have been exposed to one or more semesters of modern physics.[1] The fact that we cannot see an atom certainly does not imply that the atom does not exist. Specifically, we seek to understand the experimental results on the hydrogen atom which give the energy levels of the electron as it orbits the proton (see Fig. 8.2), which is the

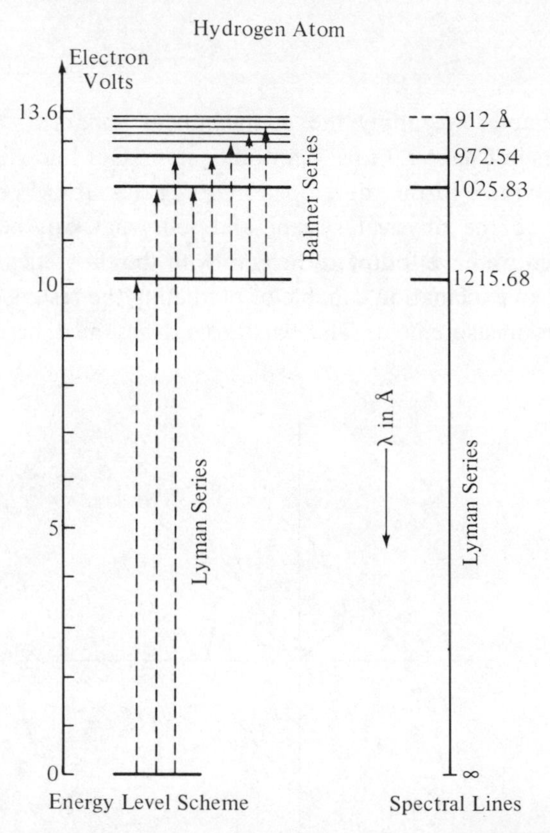

Figure 8.2 Energy Level Scheme for Hydrogen Atom and Spectral Line of Lyman Series for Hydrogen Atom.

[1] See, for example, D. C. Peaslee and H. Mueller, *Elements of Atomic Physics*, Prentice-Hall, New York (1955).

famous Lyman series, Balmer series, etc. These experimental results were the testing ground for the spherical wave theory developed during the first third of this century by Schrödinger, Bohr, Sommerfeld, and many others. The accuracy of spectrographic experiments on these energy levels is so great that a detailed theory involving a much greater refinement of the simple planetary model used here has been the subject of intense research by numerous Nobel Laureates during the second third of this century.[2] The wave theory used in this chapter is applied to a planetary model of the hydrogen atom which is made deliberately simple so that fruitful mathematical tools already developed in this book can be used.

8.1 THE HYDROGEN ATOM MODEL AS A CLASSICAL SYSTEM

Figure 8.1 is a classical mechanical concept of the hydrogen atom showing the massive proton treated as stationary and with the electron (approximately 1/1800 times smaller mass) orbiting the nuclear center. The central force binding the electron to the nucleus is the Coulomb force or potential energy, $e^2/\epsilon_0 r$. The energy, E, of the electron is assumed to be zero at infinite distance, r, so that the electron is bound to the nucleus, and the energy at finite distance, r, will be negative. The centrifugal force balanced against the Coulomb force gives for a circular orbit

$$m\omega^2 r = e^2/\epsilon_0 r^2 \qquad \text{MKS units} \tag{8.1}$$

Thus, $\omega \simeq 5 \times 10^{16}$ radians/sec gives the magnitude of angular velocity to be expected at distances $r \simeq 10^{-8}$ cm and with the values for electron charge and mass used together with the MKS unit for permittivity, ϵ_0. The linear velocity is then of the order 5×10^8 cm/sec and relativistic effects could be important in a more complete theory. Likewise the center of mass is just outside the proton and our model is recognized as a simplification, but accessible to mathematical treatment. If the electron were to change from one allowed orbit to another, it might be expected to change its energy by about 10% so that we might expect the frequency of light emitted to be some 10^{15} cycles/sec or have a wavelength of some 6000 Å (Balmer Series). The orbiting electron must have a magnetic moment given by the current in amperes multiplied by the area of the orbit's plane:

$$\beta = IA = ev\pi r^2 \simeq 10^{-23} \text{ amp-m}^2 \tag{8.2}$$

[2] An excellent review has been written by H. A. Bethe and E. E. Salpeter, *Handbuch der Physik*, Vol. 35, edited by S. Flugge, Springer, Berlin (1957).

The observed magnitude of the Bohr magneton, 9.22×10^{-24} amp-m^2 was determined by the Stern-Gerlach experiments (see Ref. 1). The ratio of the orbital magnetic moment to the orbital angular momentum, p, is

$$\frac{\beta}{p} = \frac{\pi r^2 e\nu}{m r^2 2\pi \nu} = \frac{e}{2m} \qquad \text{MKS units} \tag{8.3}$$

The magnitude to be expected for electron orbital angular momentum is 10^{-34} joule-sec.

8.2 THE HYDROGEN ATOM IN WAVE MECHANICS: Associated Legendre Polynomials

Having given the classical planetary model for the hydrogen atom and some order of magnitudes to be expected of observables, we proceed with a Schrödinger wave equation[3] for the electron (in CGS units). The space-dependent wave equation is

$$\left[\nabla^2 + \frac{2m}{\hbar^2}\left(E + \frac{e^2}{r}\right)\right]\psi = 0 \tag{8.4}$$

The physical system is characterized by a wave function, ψ, which allows us to give probability laws for the distribution of any mechanical quantity (or observable) over its possible values.

The Laplacian operator in spherical polar coordinates is

$$\nabla^2 = \frac{1}{r^2}\frac{\partial}{\partial r}\left(r^2\frac{\partial}{\partial r}\right) + \frac{1}{r^2 \sin\theta}\frac{\partial}{\partial\theta}(\sin\theta)\frac{\partial}{\partial\theta} + \frac{1}{r^2\sin^2\theta}\frac{\partial^2}{\partial\phi^2}$$

We separate into the functions

$$\psi(r, \theta, \phi) = R(r)\,\Theta(\theta)\,\Phi(\phi) \tag{8.5}$$

and substitute into Eq. 8.4 to obtain

$$\frac{1}{R^2}\frac{1}{r^2}\frac{\partial}{\partial r}\left(r^2\frac{\partial R}{\partial r}\right) + \frac{2m}{\hbar^2}\left(E + \frac{e^2}{r}\right) + \frac{1}{r^2}\left[\frac{1}{\Theta}\frac{1}{\sin\theta}\frac{\partial}{\partial\theta}(\sin\theta)\frac{\partial\Theta}{\partial\theta}\right] = -\frac{1}{r^2\sin^2\theta}\left[\frac{1}{\Phi}\frac{\partial^2\Phi}{\partial\phi^2}\right] \tag{8.6}$$

Multiplying through by $r^2 \sin^2\theta$ makes the right side independent of both r and θ so we set the right side equal to a constant, m^2, to give

[3] E. Schrödinger, *Ann. Phys.* **79**, 361 (1926).

$$\frac{d^2\Phi}{d\phi^2} = -m^2\Phi \tag{8.7}$$

This equation is identical to Eqs. 5.26 and 5.27 and has the solution

$$\Phi = e^{\pm im\phi} \qquad \text{with } m = 0, \pm 1, \pm 2, \text{etc.} \tag{8.8}$$

The wave must fit onto itself with ϕ increasing 2π radians. The wave may run clockwise or counterclockwise. In Chapter 5 for the membrane the value of $|m|$ gave the number of nodal lines the wave made through the circular membrane. In this three-dimensional case, the number $|m|$ represents the number of nodal planes through the symmetry axis.

The left side of Eq. 8.6 is also set equal to the constant, m^2, so we write (after clearing out an r^2 term),

$$\frac{1}{R^2}\frac{\partial}{\partial r}\left(r^2\frac{\partial R}{\partial r}\right) + \frac{2mr^2}{\hbar^2}\left(E + \frac{e^2}{r}\right) + \left[\frac{1}{\Theta}\frac{1}{\sin\theta}\frac{\partial}{\partial\theta}(\sin\theta)\frac{\partial\Theta}{\partial\theta}\right] = \frac{m^2}{\sin^2\theta} \tag{8.9}$$

The terms in the variable θ can be collected together on one side of Eq. 8.9 and we set the expression equal to another constant, C, to give

$$-\frac{1}{\sin\theta}\frac{\partial}{\partial\theta}(\sin\theta)\frac{\partial\Theta}{\partial\theta} + \frac{m^2\Theta}{\sin^2\theta} = C\Theta \tag{8.10}$$

or we can write

$$\frac{1}{\sin\theta}\frac{d}{d\theta}\left(\sin\theta\frac{d\Theta}{d\theta}\right) + \left(C - \frac{m^2}{\sin^2\theta}\right)\Theta = 0 \tag{8.10$'$}$$

Let $\cos\theta = \mu$ and $d\mu = -\sin\theta\, d\theta$,

$$\sin\theta\frac{d}{d\theta} = \sin^2\frac{d}{\sin\theta\, d\theta} = -(1 - \mu^2)\frac{d}{d\mu}$$

Then Eq. 8.10$'$ becomes

$$(1 - \mu^2)\frac{d^2\Theta}{d\mu^2} - 2\mu\frac{d\Theta}{d\mu} + \left(C - \frac{m^2}{1 - \mu^2}\right)\Theta = 0 \tag{8.11}$$

If $m = 0$, then the above equation is just like Eq. 7.13$'$, and the discussion of the solution to that equation showed us that

$$C = l(l + 1) \qquad \text{with } l = 0, 1, 2, 3, \text{etc.} \tag{8.12}$$

For Eq. 7.13$'$ we had solutions which were Legendre polynomials and we therefore changed the symbol $\Theta_l(\mu) \equiv P_l(\mu)$. Now we seek a solution to an Associated Legendre differential equation, Eq. 8.11. $\Theta_l^m(\mu) \equiv P_l^m(\mu)$ denotes the solution in Associated Legendre polynomials whose analytic expression comes from the solution of the equation:

$$(1-\mu^2)\frac{d^2P_l^m(\mu)}{d\mu^2} - 2\mu\frac{dP_l^m(\mu)}{d\mu} + \left[l(l+1) - \frac{m^2}{1-\mu^2}\right]P_l^m(\mu) = 0 \tag{8.13}$$

We transform this last equation by substituting into it

$$P_l^m(\mu) = (1-\mu)^{m/2}P \tag{8.14}$$

and

$$\frac{d}{d\mu}(1-\mu^2)^{m/2}P = \left[\frac{m}{2}(1-\mu^2)^{(m/2)-1}(-2\mu)\right]P + (1-\mu^2)^{m/2}\frac{dP}{d\mu} \tag{8.15}$$

and so on, to give after a little algebra

$$(1-\mu^2)\frac{d^2P}{d\mu^2} - 2(m+1)\mu\frac{dP}{d\mu} + [l(l+1) - m(m+1)]P = 0 \tag{8.16}$$

This last equation is not quite the simple differential equation, Eq. 7.13′, because of the extra constants $(m+1)$ and m. But we recognize[4] that P is just the mth derivative of the familiar Legendre function, called $P_l(\mu)$, that does satisfy the Legendre differential equation, Eq. 7.13′; i.e.,

$$P = \frac{d^mP_l(\mu)}{d\mu^m} \tag{8.17}$$

The complete solution to Eq. 8.13 is

$$P_l^m(\mu) = (1-\mu^2)^{m/2}\frac{d^mP_l(\mu)}{d\mu^m} \tag{8.18}$$

A few of the terms, called surface zonal harmonics, are as follows:

$P_0^0 = 1$ because $P_0(\cos\theta) = 1$
and with $m = 0$, no differentiation

$P_1^0 = \cos\theta,\ P_1^1 = \sin\theta$ because $(1-\mu^2)^{1/2} = \sin\theta$
and $d\mu/d\mu = 1$

$P_2^0 = \frac{1}{2}(3\cos^2\theta - 1),\ P_2^1 = 3\sin\theta\cos\theta,\ P_2^2 = 3\sin^2\theta$

Obviously, $P_l^m = 0$ for $m > l$ because the order of differentiation must not

[4]See, for example, A. Sommerfeld, *Partial Differential Equations in Physics*, 4th printing, page 128, Academic Press, New York, (1967).

be greater than the degree of the differentiated polynomial. Returning to the original problem, the solution to Eq. 8.4 can be written

$$\psi(r,\theta,\phi) = R(r)\, P_l^m(\cos\theta)\, e^{im\phi} \tag{8.19}$$

The terms $P_l^m(\cos\theta)\, e^{im\phi}$ together form the (unnormalized) spherical harmonic. In Eq. 8.8 we allowed negative values of m for waves running clockwise in ϕ so that $|m| \leq l$ or we have a total of $2l + 1$ values of $m = -l, \cdots, l-1, l$. These integer m values are a measure of the orbital angular momentum in the z-direction. They are called "magnetic quantum numbers" because they appear in a theory of the experiment in which an external magnetic field is applied along the z-direction (Zeeman effect). The angular momentum about the z-axis is calculated from the following operation on the wave function:

$$-i\hbar\left(\frac{x\partial}{\partial y} - \frac{y\partial}{\partial x}\right)\psi = -i\hbar\frac{\partial}{\partial\phi}\psi \text{ in polar coordinates.}$$

Using Eq. 8.19, the differentiation gives

$$-i\hbar\frac{\partial}{\partial\phi}[R(r)P_l^m(\cos\theta)e^{im\phi}] = m\hbar\psi \tag{8.20}$$

and we obtain the characteristic values, $m\hbar$, of the angular momentum operator. In order to insure the probability meaning of the wave function, $\int \psi^*\psi\, dx\, dy\, dz = 1$, we must suitably normalize each part of the total wave function in Eq. 8.19. When this is done the result is that the expectation value of the angular momentum about the z-axis is

$$\bar{p}_z = \int \psi^*\left(-i\hbar\frac{\partial}{\partial\phi}\right)\psi\, dx\, dy\, dz = m\hbar \tag{8.21}$$

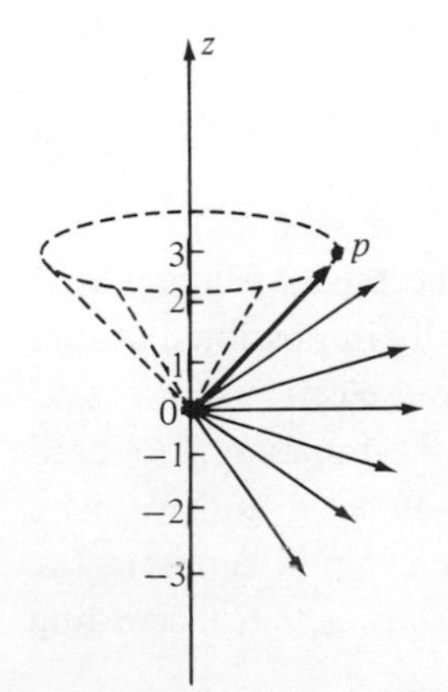

Figure 8.3 Projections of Total Angular Momentum on z-axis are quantized to values $m\hbar$ where m can take values such as −3, −2, −1, 0, 1, 2, and 3. The total angular momentum vector precesses about the z-axis to sweep out a cone as illustrated by the m = 3 case.

The important point about this part of the theory is that the z-component of p is quantized. This was not to be expected from the classical planetary model. With $\hbar = 1.05 \times 10^{-34}$ joule-sec, the value for $m = 1$ is in agreement with $mr^2\omega$ computed from Eq. 8.1 or 8.3. The space quantization does show up in the Stern-Gerlach experiments and we show graphically in Fig. 8.3 the quantization of p_z and indicate that the total momentum, p, may be thought to precess about the z-axis without having any fixed position in space.

8.3 RADIAL WAVE FUNCTION FOR HYDROGEN ATOM: Associated Laguerre Polynomial

Returning to Eq. 8.6, we now write the radial part of the wave equation as the expression

$$\frac{d^2R}{dr^2} + \frac{2}{r}\frac{dR}{dr} + \left\{\frac{2m}{\hbar^2}\left(E - \frac{e^2}{r}\right) - \frac{l(l+1)}{r^2}\right\}R = 0 \tag{8.22}$$

We recognize that this last equation is very similar to the radial wave equation, Eq. 7.17, of Chapter 7. At great distance the coefficient of R in Eq. 8.22 approaches just $(2mE/\hbar^2)^{1/2}$ because terms in $1/r$ and $1/r^2$ become increasingly small. But for a bound electron, E is negative so we have

$$k = \pm\sqrt{\frac{-2m\,|E|}{\hbar^2}} = \pm i\sqrt{\frac{2m\,|E|}{\hbar^2}}$$

which is pure imaginary. At large r Eq. 8.22 reduces to a very familiar differential equation

$$\frac{d^2R}{dr^2} + k^2R = 0 \tag{8.23}$$

The familiar solution is

$$R = Ae^{\pm ikr} = A\exp \pm r\left(\frac{2m\,|E|}{\hbar^2}\right)^{1/2} \tag{8.24}$$

which is not a plane wave oscillating about the r-axis because k is imaginary. We exclude the plus sign in the exponential as physically impossible because the wave function must vanish at large r for a captured electron.

In Fig. 8.4 we have represented by a dashed line the function r^2R^2 used in the expectation value of the electron density at distance r, and we have drawn a full line beyond some large distance r_0 which might represent the true situation for a valid Eq. 8.24. What we notice at once is that, according

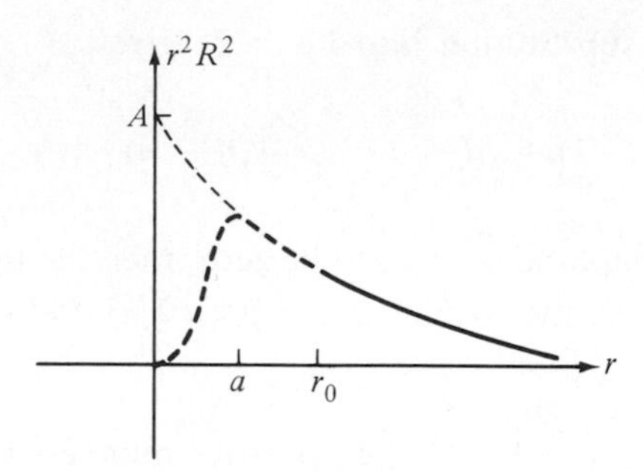

Figure 8.4 Radial Wave Function for Hydrogen Atom. The probability density peaks up at distance *a* called the Bohr radius. The approximation at large distance *r* is considered valid beyond r_0.

to this figure, the electron is distributed spherically over the region of space out past r_0 where the theory is considered reliable. It appears that we do *not* have an electron at a point in space as shown in Fig. 8.1, but we have a charge cloud, $e|\psi|^2$, smeared around the proton. This corresponds to a point-like electron whose position is not exactly known and we are *not* to understand the charge cloud as a smeared-out electron. The cloud has no sharp outer boundary, but falls off rapidly as $r \longrightarrow \infty$.

Let us now investigate the dependence of R on r at small distances. We expect peaks and valleys from the radial part of the wave equation in the way Eq. 7.21 gave a spherical Bessel function for dipole and multi-pole pressure radiation. Following Stokes, we might take $R = A(r)\exp(-kr)$ and seek a polynomial for $A(r)$ as was done in Chapter 7. We delay this proper attack for a special solution, or trick, so that we may reveal some physical properties as quickly as possible.

Equation 8.22 for $R(r)$ may be transformed, as we did Eq. 7.17 in the last chapter, with the following symbols and parameters:

$$z = kr \qquad \text{with } -k^2 = 2mE/\hbar^2$$

$$\eta = me^2/\hbar^2 k$$

$$R(r) = \chi(r)/r$$

Equation 8.22 now becomes

$$\frac{d^2\chi}{dz^2} + \left[-1 - \frac{l(l+1)}{z^2} - \frac{2\eta}{z}\right]\chi = 0 \tag{8.25}$$

Consider the special solution

$$\chi = Az^n e^{-z} \tag{8.26}$$

Differentiation and substitution into Eq. 8.25 gives

$$\left[\frac{1}{z^2}\{n(n-1)-l(l+1)\}-\frac{2}{z}(\eta-n)\right]Az^n e^{-z}=0 \qquad (8.27)$$

If A, the arbitrary amplitude, is not to be zero, then the terms in the bracket [] must be zero and so the coefficients of both $1/z^2$ and of $1/z$ must vanish. We therefore obtain

$\eta = n$ where $n = 1, 2, 3$, etc., positive numbers only to keep χ finite at $z = 0$

and $$n = l + 1$$

Thus the radial quantum number as it appears here is one greater than the orbital number l. Later we must qualify this result and call n the total quantum number and show that it is not uniquely determined by l. Putting in the value of η and k, we have

$$\eta = n = me^2/\hbar^2 k = me^2/\hbar^2(-2mE/\hbar^2)^{1/2} \qquad (8.28)$$

or $E = E_n = -me^4/2\hbar^2 n^2$ the minus sign means work must be done to remove the captured electron to $r = \infty$

This formula was given by N. Bohr,[5] and n is called the Bohr quantum number. The observed frequency of a spectral line is obtained by the passage of the electron from an initial excited state E_1 to a final lower state of energy E_2 so that we have

$$h\nu = E_2 - E_1 = \frac{2\pi^2 me^4}{h^2}\left[\frac{1}{n_2^2}-\frac{1}{n_1^2}\right] \qquad (8.29)$$

or $$\nu = R\left[\frac{1}{n_2^2}-\frac{1}{n_1^2}\right] \qquad (8.30)$$ with $R = 2\pi^2 me^4/h^3$ the Rydberg frequency

The Lyman series is obtained if the final state is the ground state of the hydrogen atom $n_2 = 1$. The Balmer series corresponds to the passage to the final state $n_2 = 2$. We write these as follows:

$$\nu = R\left(\frac{1}{1^2}-\frac{1}{n^2}\right) \qquad n = 2, 3, 4, \text{etc.} \qquad \text{Lyman Series} \qquad (8.31)$$

$$\nu = R\left(\frac{1}{2^2}-\frac{1}{n^2}\right) \qquad n = 3, 4, 5, \text{etc.} \qquad \text{Balmer Series} \qquad (8.32)$$

[5] *Phil. Mag.*, **26**, 1 (1913).

The theory does agree with the experimental facts shown in Fig. 8.2. Since the trial solution Eq. 8.26 seems to fit the experimental facts as given by Eqs. 8.31 and 8.32, we may wish to let well enough alone. The characteristic function has an r dependence that makes it vanish not only as $r \longrightarrow \infty$ but also as $r \longrightarrow 0$. Physically, this is very appealing for a stable hydrogen atom. The quantity $k = me^2/\hbar^2 n$ has the dimensions of $1/r$, and with the Bohr quantum number $n = 1$ we have $a = 1/k = \hbar^2/me^2 = 1/2 \times 10^{-8}$ cm, called the Bohr hydrogen radius. The electron has an appreciable probability density at $r = a$, but is smeared off to either side of a as shown in Fig. 8.4. The order of magnitude calculations using $r = 10^{-8}$ cm made at the beginning of this chapter (Eqs. 8.1, 8.2, and 8.3) are reasonable.

Arnold Sommerfeld, who contributed so much to this theory himself, has given an excellent summary (Ref. 4 this chapter and also Ref. 2) concerning the solution of the radial part of the wave equation, and for us not to include the radial solution would be like omitting the last movement of a symphony.

As promised, we finally consider the solution of Eq. 8.22 to be $R(r) = A(r) \exp(-kr)$ and substitute back to find that $A(r)$ must satisfy the equation

$$\frac{d^2A(r)}{dr^2} + \left(\frac{2}{r} - 2k\right)\frac{dA(r)}{dr} + \left[\frac{(2n-2)k}{r} - \frac{l(l+1)}{r^2}\right]A(r) = 0 \qquad (8.33)$$

The symbols n and k have been defined in the paragraphs above. Transforming to the variable $\rho = 2kr$ we obtain

$$\frac{d^2A(\rho)}{d\rho^2} + \left(\frac{2}{\rho} - 1\right)\frac{dA(\rho)}{d\rho} + \left[\frac{n-1}{\rho} - \frac{l(l+1)}{\rho^2}\right]A(\rho) = 0 \qquad (8.34)$$

which is in the form used by Sommerfeld and other well known texts[6] on quantum mechanics. The polynomial we seek is written

$$A(\rho) = \rho^{\lambda}(a_0 + a_1\rho + a_2\rho + \cdots a_p\rho^p + \cdots) \qquad (8.35)$$

This is a procedure exactly like Eq. 7.19 on pressure waves used to solve Eq. 7.18′. Substitution of Eq. 8.35 and its derivatives into Eq. 8.34 allows us to determine that

$$\lambda(\lambda + 1) = l(l + 1) \qquad (8.36)$$

or $\quad \lambda = l \quad$ the other root, $\lambda = -l - 1$ is excluded in order to make $A(\rho)$ finite at $\rho = 0$. (8.37)

[6]For example, L. I. Schiff, *Quantum Mechanics*, page 83, McGraw-Hill Book Co., New York (1949).

As in Chapter 7 where these techniques were used, the recursion formula for the coefficients turns out to be

$$a_{p+1}[(\lambda + p + 1)(\lambda + p) + 2(\lambda + p + 1) - l(l + 1)] \\ = -a_p[n - 1 - \lambda - p] \tag{8.38}$$

The polynomial may break off such that a_{p+1} and all subsequent terms vanish if we make the coefficient of a_p in the above equation vanish by setting

$$n = p + \lambda + 1 = p + l + 1 \tag{8.39}$$

If the polynomial were not to break off, then the series would diverge for $\rho \longrightarrow \infty$. The integer p takes on a special meaning for the physical interpretation of the radial part of the wave equation. Comparing Eq. 8.39 with the value $n = l + 1$ found for the very special solution Eq. 8.26, we see that we had selected a possible integer value $p = 0$ for that trial solution. Thus the total quantum number, n, depends on the value of the quantum number p *and* the quantum number l. We have for the total wave function, ψ, a total of three independent quantum numbers: m, l, and p. For three-dimensional space this is essential. The quantum number p is sometimes written n_r as a symbol for radial quantum number. It gives the number of zeros of $R(r)$ as $r \longrightarrow \infty$. What is the nature of this new polynomial? From analogy to the Legendre polynomial, what differential equation does this polynomial satisfy? From Eq. 8.35 we define the new polynomial by the symbol as follows:

$$A(\rho) = \rho^l(a_0 + a_1\rho + a_2\rho^2 + \cdots a_p\rho^p) \equiv \rho^l L(\rho) \tag{8.40}$$

We understand that the series stops at a value $p = n - l - 1$. If we substitute this expression and its derivatives into Eq. 8.34 we obtain (with use of Eqs. 8.37 and 8.39)

$$\rho\frac{d^2L}{d\rho^2} + [2(l + 1) - \rho]\frac{dL}{d\rho} + (n - l - 1)L = 0 \tag{8.41}$$

We wish to show that one solution to this last equation is the Associated Laguerre polynomial. This last differential equation strongly resembles the Laguerre differential equation with the integer $\mu = n + l$. The Laguerre differential equation is

$$\rho\frac{d^2L_\mu}{d\rho^2} + [1 - \rho]\frac{dL_\mu}{d\rho} + \mu L_\mu = 0 \tag{8.42}$$

The solution of this last equation is labeled $L_\mu(\rho)$ and is the Laguerre polynomial of degree μ. Just as in Eq. 8.16 there was a strong resemblance to the Legendre differential equation, so by analogy to Eq. 8.17 we now recognize that the Associated Laguerre polynomial is the solution of Eq. 8.41 and is written

$$L \equiv L_{n+l}^{2l+1} = \frac{d^{2l+1}L_{\mu}(\rho)}{d\rho^{2l+1}} = \frac{d^{2l+1}L_{n+l}(\rho)}{d\rho^{2l+1}} \tag{8.43}$$

The solution, L, to Eq. 8.41 is obtained through Eq. 8.43: i.e., the $(2l + 1)$-fold differentiation of the solution to Eq. 8.42. We have labeled the solution $L_{n+l}^{2l+1}(\rho)$ and for a few examples of this radial function we refer to Refs. 4 and 2 of this chapter. Certain values of $\rho = 2kr$ make the Associated Laguerre polynomial go to zero, just as in earlier chapters the Bessel function crossed the radial axis. There are $n - l - 1$ zeros of the radial function. We summarize this theory by writing the characteristic wave function for the hydrogen atom's single electron as

$$\psi = A\rho^{l}e^{-\rho/2}\frac{d^{2l+1}L_{n+l}(\rho)}{d\rho^{2l+1}}P_{l}^{m}(\cos\theta)e^{im\phi} \tag{8.44}$$

In terms of the variable, $\mathbf{k}r$, we may write the wave function as

$$\psi(r,\theta,\phi) = A(2kr)^{l}e^{-\mathbf{k}r}\frac{d^{2l+1}L_{n+l}(2\mathbf{k}r)}{d(2kr)^{2l+1}}P_{l}^{m}(\cos\theta)e^{im\phi} \tag{8.44'}$$

To discuss these matters further or to take an improved (more realistic) model would be to expand this study into the details of a proper book on quantum mechanics. There would be the electron spin and many remarkable details to add to what has been said here. Our aim has been to carry the reader to this threshold and to show that the elementary wave theory of the hydrogen atom is only a slight extension of the mathematical tools acquired in classical systems. A certain unity of viewpoint is brought about by this presentation of wave mechanics.

8.4 THE TWO-BODY PROBLEM: Phase Shift

Figure 8.5 shows the scheme of two particles at vector positions $\mathbf{r}_1$ and $\mathbf{r}_2$ having momenta $\mathbf{p}_1$ and $\mathbf{p}_2$ and drawn to indicate that they will pass near to one another. To be definite, let us take these particles as identical argon atoms of mass, m, and having a weak potential of interaction like Eq. 6.1 for the Lennard-Jones potential

$$V(r) = 4\epsilon\left[\left(\frac{\sigma}{r}\right)^{12} - \left(\frac{\sigma}{r}\right)^{6}\right]$$

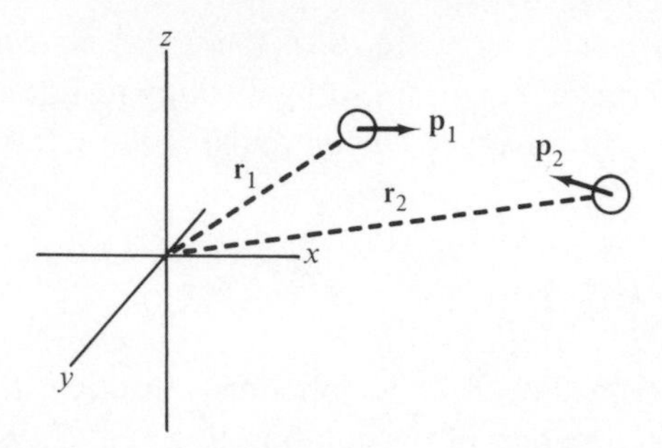

Figure 8.5 Schematic Drawing for Two-Body Problem with Lennard-Jones Potential and Positive Energy.

where r is the distance between the particles $r = r_1 - r_2$. At distances $r > 3\sigma$ the potential of attraction is so weak that the particles are essentially free particles beyond that distance. Close to one another, the two interacting particles are moving in a centrally symmetric field and most of our mathematical tools in the past sections of this chapter and in Chapter 7 will be useful again. The importance of this two-body problem can be illustrated by pointing out that deviations from a perfect gas law $PV = RT$ can be expected when pairs of atoms having a potential $V(r)$ are in near collision distance. The theory given here has been used in studies on gases by many scientists; for example, by J. De Boer[7] and by the two Nobel Laureates, T. D. Lee and C. N. Yang.[8]

The problem is unlike the hydrogen atom just discussed in that the energy is positive and we do not have discrete energy states characteristic of a bound particle problem. The two particles in Fig. 8.5 do have angular momentum and radial momentum which is made especially clear when expressed in a center of mass coordinate system. The customary Schrödinger equation for two particles of identical mass is

$$\left[-\frac{\hbar^2}{2m}\nabla_1^2 - \frac{\hbar^2}{2m}\nabla_2^2 + V(r)\right]\Psi = E\Psi \tag{8.45}$$

where ∇_1^2 and ∇_2^2 are the Laplacian operators with respect to the coordinates of the two particles. If the potential $V(r) = 0$, then the wave function for the two free atoms is the product of plane waves:

$$\Psi = \frac{1}{V}e^{(i/\hbar)p_1r_1} \cdot e^{(i/\hbar)p_2r_2} \tag{8.46}$$

[7]J. De Boer, "Contributions to the Theory of Compressed Gases," Ph.D. Thesis, Amsterdam, 1940.

[8]T. D. Lee and C. N. Yang, *Phys. Rev.* **113**, 1165, 1959.

and the coefficient insures that the two particles in a volume, V, will give $\int\int |\Psi|^2 \, dV_1 \, dV_2 = 1$. No special restrictions of the symmetry of the wave function is imposed at this time, and for many applications none are needed.

Introducing the new variables $r = r_1 - r_2$ and $R = \frac{1}{2}(r_1 + r_2)$ the Schrödinger equation becomes

$$\left[-\frac{\hbar^2}{4m}\nabla_R^2 - \frac{\hbar^2}{2\mu}\nabla_r^2 + V(r)\right]\Psi = E\Psi \tag{8.47}$$

where $\mu = m/2 = m_1 m_2/(m_1 + m_2)$ is the reduced mass. We expect a solution

$$\Psi(r_1 r_2) = \Phi(R)\,\psi(r) \tag{8.48}$$

where $\Phi(R)$ is just the plane wave of a free particle of mass $2m$. Our interest centers on the wave function $\psi(r)$ in the theory which follows.

The wave equation in the relative coordinates is

$$\nabla^2\psi + \frac{2\mu}{\hbar^2}[E - V(r)]\psi = 0 \tag{8.49}$$

which is similar to Eq. 8.4, and the Laplacian in radial vector distance r can be replaced by the Laplacian in spherical polar coordinates just as was done for Eq. 8.4. We have characteristic functions in ϕ and in θ exactly like Eqs. 8.8 and 8.18 so that we have a solution

$$\psi(r,\theta,\phi) = R(r)P_l^m(\cos\theta)e^{im\phi} = R(r)Y_{lm}(\theta,\phi) \tag{8.50}$$

The radial function, $R(r)$, satisfies the equation

$$\frac{1}{r^2}\frac{d}{dr}\left(r^2\frac{dR}{dr}\right) - \frac{l(l+1)}{r^2}R + \frac{2\mu}{\hbar^2}(E - V(r))R = 0 \tag{8.51}$$

which is just like Eq. 8.22 except that now E is positive. We have learned that outgoing spherical waves fall off as $1/r$ in all physics, so following the procedure leading to Eq. 8.25 we substitute $R(r) = \chi(r)/r$ and obtain the equation

$$\frac{d^2\chi}{dr^2} + \left[\frac{2m}{\hbar^2}(E - V(r)) - \frac{l(l+1)}{r^2}\right]\chi = 0 \tag{8.52}$$

Next we substitute the customary variable, $z = kr$, where $k^2 = 2mE/\hbar^2$:

$$\frac{d^2\chi}{dz^2} + \left[1 - \frac{l(l+1)}{z^2} - \frac{2m}{\hbar^2 k^2}V(r)\right]\chi = 0 \tag{8.53}$$

We now wish to discuss the solution of this equation, and with the Lennard-Jones potential for $V(r)$ this is no simple task. We must resort to numerical integration. But we can bring out some interesting features by comparing the true solution with that which comes from setting $V(r) = 0$. Of course, for completely free particles $V(r) = 0$ and the solution in spherical coordinates coming from Eq. 8.53 must be the same as the plane wave solution of Eq. 8.46. For $V(r) = 0$, Eq. 8.53 becomes

$$\frac{d^2\chi}{dz^2} + \left[1 - \frac{l(l+1)}{z^2}\right]\chi = 0 \tag{8.54}$$

Let the solution be

$$\chi = a_0 z^s + a_1 z^{s+1} + a_2 z^{s+2} + \cdots$$

By differentiation and setting to zero the coefficient of z^{s-2}, z^{s-1}, etc., we have

$$s = l + 1 \qquad \text{or} \qquad s = -l$$

for arbitrary value of a_0. Furthermore, we find that $a_1 = 0$, $a_3 = 0$, etc. The even-numbered coefficients have a recursion relation, e.g.,

$$a_2 = -a_0[1/(l+2)(l+3) + 1]$$

The series for χ is recognized as a Bessel function for the value $s = l + 1$ which is

$$\chi = a_0\sqrt{(\pi z/2)}J_{l+1/2}(z) \tag{8.55}$$

We exclude the series generated by $s = -l$ in order to keep $R(r)$ finite. For very small values of r we can take the first term of the power series solution of Eq. 8.54 and have $R(r) = a_0 r^l$. The asymptotic value of Eq. 8.55 for $r \longrightarrow \infty$ is

$$\chi \equiv A \sin\left(\mathbf{k}r - \frac{l\pi}{2}\right) \tag{8.56}$$

At distances $r > 3\sigma$ we may consider the Lennard-Jones potential as zero, $V(r) = 0$, so that Eq. 8.54 is again valid for those values of $r > 3\sigma$. But we must not exclude the series $s = -l$, and so we have two possible series given by

$$\chi = A_l(k)\sqrt{(\pi kr/2)}J_{l+1/2}(\mathbf{k}r) + B_l(k)\sqrt{(\pi kr/2)}J_{-l-1/2}(\mathbf{k}r) \tag{8.57}$$

The solution must have coefficients $A_l(k)$ and $B_l(k)$ such that the real wave function, χ, which satisfies Eq. 8.53 and its derivative with respect to r be continuous at $r = 3\sigma$ to the approximation wave function given by Eq. 8.57. For large values of r in Eq. 8.57 the Bessel functions can be approached by a sine respectively cosine function and, taken together, these two asymptotic solutions give

$$\chi \longrightarrow (A^2 + B^2)^{1/2} \sin\left(\mathbf{k}r - \frac{l\pi}{2} + \eta_l(\mathbf{k})\right) \tag{8.58}$$

where the phase shift, $\eta_l(\mathbf{k})$, results from the existence of the potential $V(r)$. Lee and Yang (Ref. 8) have used Eq. 8.58 in their discussion of the so-called "hard sphere" potential and they give reasons for neglecting $\eta_l(k)$ in that example. De Boer (Ref. 7) has calculated the phase shift numerically for helium atoms at low temperatures. With this he calculated the correction to the ideal gas law. As in many problems of the physics of actual systems, we cannot have a neat analytic solution to an equation like Eq. 8.53. Numerical integration with the help of computers is not very difficult. The wave theory discussed in this last section is the basic theory of atomic scattering, and it is also the theory of the central force problem.

PROBLEMS

1. The electron cloud surrounding the nucleus can be treated as being bound as though with a linear restoring force, $-kx$. The distance, x, is the amount by which the spherical electron cloud has its center shifted away from the positive charge nucleus. The potential is $V = 1/2kx^2$ and the Schrödinger equation is

$$\frac{d^2\psi}{dx^2} + \frac{2m}{\hbar^2}(E - \tfrac{1}{2}kx^2)\psi = 0$$

Show that $\psi = xe^{-cx^2}$ satisfies this equation and compute the energy E_1 for this state.

2. In the above harmonic oscillator problem, discuss the solution at large distance x such that the wave equation can neglect the energy E and be written

$$\frac{d^2\psi}{dx^2} + \frac{2m}{\hbar^2}(-\tfrac{1}{2}kx^2)\psi = 0$$

3. A diatomic molecule can be taken as a rigid rotor with mass points m_1 and m_2 located a distance r apart. The moment of inertia about the important axis of rotation is

$$\left(\frac{m_1 m_2}{m_1 + m_2}\right) r = \mu r^2$$

Thus the rotor can be treated as an effective mass μ at a distance r and the Schrödinger equation written

$$\nabla^2\psi + \left(\frac{2\mu}{\hbar^2}E\right)\psi = 0$$

Express this equation in spherical coordinates and show that for the rigid rotor (no dependence on r) the wave solution is

$$\psi = NP_J^m(\cos\theta)\, e^{im\phi}$$

with energy values

$$E_J = \frac{\hbar^2 J(J+1)}{2\mu r^2}$$

Chapter 9

ELEMENTARY WAVE THEORY OF SCATTERING

In this final chapter we discuss scattering of waves as a topic of basic importance to many areas of physical science. Much of the approach to the theory of scattering of an incident plane wave by an atom or nucleus may be said to have its beginning in the work by Lord Rayleigh.[1] Rayleigh worked out the scattering of plane waves of sound by a rigid spherical object. The scattering of light (Rayleigh, 1871) has also been of great importance to our knowledge of matter. The theory of scattering of electron beams was introduced by Faxén and Holtsmark[2] during the most exciting development of quantum wave mechanics. One of the first books on the quantum theory of scattering of a beam of particles was by N. F. Mott and H. S. W. Massey.[3] This book reviewed the original contributions of scientists such as Bohr, Born, Dirac, Rutherford, Wigner, Breit, Bethe, Bloch, Sommerfeld, Wentzel, Kramers, and Brillouin. The scattering of X-rays and of neutrons is also of enormous importance to the knowledge of the structure of matter. Our modest aim in this chapter is to touch some of these topics by way of making use of the mathematical tools and physical ideas already developed in this book. In this way we are introduced to the subject of scattering and present all the rich background taken for granted by those who write papers and books on the modern theory of wave scattering.

[1] Lord Rayleigh, *The Theory of Sound*, Vol. 2, page 272, Dover Publications, New York (1945).

[2] H. Faxén and J. Holtsmark, *Zeit. f. Physik*, **45**, 307 (1927).

[3] N. F. Mott and H. S. W. Massey, *The Theory of Atomic Collisions*, Oxford Press (1933); 3rd edition (1965).

9.1 ELASTIC COLLISIONS (Electron Scattering): Born Formula

The incoming plane wave, a free particle, is described by $\psi = e^{ikz}$ where $\mathbf{k} = mv/\hbar$ for the wave vector of a particle. The scattered wave is described at great distance from the scattering center by an outgoing spherical wave $\approx e^{ikr}/r$. Experimentally, we measure the intensity of the scattered particles at an angle and the detector has some finite element of solid angle. The discussion at the end of Chapter 8 showed the form of the asymptotic solution of the two-body scattering problem with potential between the particles $V(r)$ to be

$$\psi \approx \sum_l A_l P_l(\cos\theta)\frac{\sin(\mathbf{k}r - (l\pi/2) + \eta_l(k))}{kr} \tag{9.1}$$

We recall that Eq. 9.1 is the wave made up of both incoming and outgoing scattered particles. On the other hand, the incoming free particle can have the plane wave represented by a series of harmonics (see Eq. 7.32). The asymptotic value is

$$e^{i\mathbf{k}z} \simeq \sum_l i^l(2l+1)P_l(\cos\theta)\frac{\sin(\mathbf{k}r - l\pi/2)}{\mathbf{k}r} \tag{9.2}$$

The coefficient, $i^l(2l + 1)$, was determined by matching expansion terms of the left side with the right side. The axial symmetry of the problem requires the quantum number $m = 0$ in both of the above equations. The difference between Eqs. 9.1 and 9.2 must represent the outgoing wave

$$\psi - e^{i\mathbf{k}z} \simeq \sum\nolimits_l \frac{1}{kr}P_l(\cos\theta)\left\{A_l\sin\left(\mathbf{k}r - \frac{l\pi}{2} + \eta_l(\mathbf{k})\right)\right.$$
$$\left. - i^l(2l+1)\sin\left(\mathbf{k}r - \frac{l\pi}{2}\right)\right\} \tag{9.3}$$

The term in the braces, { }, can be written with equivalent exponential functions instead of the sine functions (see Eq. 2.12′) as follows:

$$\{\quad\} = \frac{1}{2i}e^{i(\mathbf{k}r - l\pi/2)}[A_l e^{i\eta_l} - i^l(2l+1)]$$
$$-\frac{1}{2i}e^{-i(\mathbf{k}r - l\pi/2)}[A_l e^{-i\eta_l} - i^l(2l+1)] \tag{9.4}$$

We have to determine the coefficients A_l. In order that Eq. 9.4 put into Eq. 9.3 may make the resulting outgoing wave approach the expression $e^{i\mathbf{k}r}/r$, we must select the values of the coefficients such that

$$A^l = (2l + 1)i^l e^{i\eta_l} \tag{9.5}$$

In this way the second term on the right of Eq. 9.4 will vanish. Thus the asymptotic form of the scattered wave $f(\theta)(e^{ikr}/r)$ is obtained with the θ dependent function; i.e., the scattering angle function, given by

$$f(\theta) = \frac{1}{2ik}\sum_{l=0}^{\infty}(2l + 1)\,|\,e^{2i\eta_l} - 1\,|\,P_l(\cos\theta) \tag{9.6}$$

If $\eta_l \ll 1$, we have, by expanding the exponential and taking only the first and second terms,

$$f(\theta) = \frac{1}{k}\sum_{l=0}^{\infty}(2l + 1)\eta_l P_l(\cos\theta) \qquad \text{called the "Born formula"} \tag{9.6'}$$

Figure 9.1 shows the scattering of an incoming particle by a target particle which may be considered at rest. The probability per second that

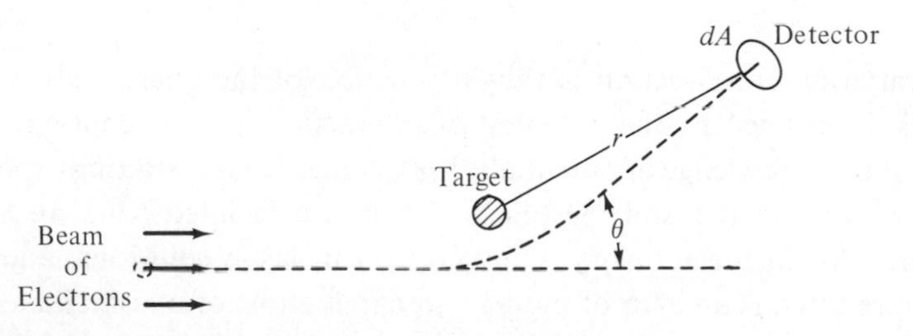

Figure 9.1 Schematic Diagram of Scattering by Beam of Particles from Atomic Target.

the scattered particle will pass out through a surface element dA located a distance r from the scattering center is

$$dA\left\{v\left[f(\theta)\frac{e^{ikr}}{r}\right]^*\left[f(\theta)\frac{e^{ikr}}{r}\right]\right\} = \frac{v}{r^2}\,|\,f(\theta)\,|^2 dA = v\,|\,f(\theta)\,|^2 d\Omega$$

where $d\Omega$ is the solid angle and v is the particle velocity. The probability density for the incident wave made up of a beam of particles whose cross section of the beam is 1 cm² is

$$v[e^{ikz}]^*[e^{ikz}] = v$$

The ratio of these probabilities is

$$d\sigma = |\,f(\theta)\,|^2\, d\Omega$$

or we may write

$$d\sigma = |f(\theta)|^2 2\pi \sin\theta \, d\theta \tag{9.7}$$

which is called the effective cross section for scattering.

The quantity $|f(\theta)|^2$ is called the scattering amplitude. The total effective cross section for scattering is

$$\sigma = 2\pi \int_0^\pi |f(\theta)|^2 \sin\theta \, d\theta \tag{9.8}$$

Before using Eq. 9.6 in this last expression, we note that the orthogonal properties of the Legendre polynomials give the relation

$$\int_0^\pi P_l^2(\cos\theta) \sin\theta \, d\theta = \frac{2}{2l+1} \tag{9.9}$$

Thus the total effective cross section for scattering is

$$\sigma = \frac{4\pi}{k^2} \sum_{l=0}^{\infty} (2l+1) \sin^2 \eta_l(\mathbf{k}) \tag{9.8$'$}$$

The scattering cross section is thus a function of the phase shift which, in turn, is determined by the potential of interaction, the incident energy, etc. Much of our knowledge of potentials, $V(r)$, comes from scattering experiments where σ is measured and the phase, $\eta_l(k)$, is calculated from an assumed potential. Much more theory is involved in inelastic collisions, and indeed this entire topic is an area of modern research effort of considerable magnitude. Finally, we remark that electron scattering with a beam of very high energy electrons has been used as a probe of nuclear and nucleon structure by the Nobel Laureate, Robert Hofstadter. The elementary theory of Faxén and Holtsmark on electron scattering which was reviewed in the above paragraphs is not an adequate approximation to the theory required for an understanding of Hofstadter's experiments.

9.2 THE SCATTERING OF PHOTONS BY ATOMS AND MOLECULES

Light waves treated as electromagnetic waves propagating in a transparent medium are familiar to beginning students of science. Wave equations in electric and magnetic field vectors are derived from Maxwell's equations. Reflection, refraction, and dispersion are phenomena explained in early science courses by use of dielectric and magnetic properties of the media which the incident beam enters. The incident wave, through its electric and magnetic field, may drive atomic electric and magnetic dipoles (or induce their existence in the first place). The magnetic property is very important for radio waves in substances like ferrite. But for the optical and X-ray region of interest in the present discussion, the electric field and the motion of electrons on atoms in the medium caused by this field are of primary importance. The force, eE, on an electron of charge e is enormous compared to the magnetic force. The motion of the charges in the medium results in the emission of scattered waves; i.e., each atom is an emitter of a weak secondary radiation. The phenomenon of diffraction results from interference effects among wave trains originating from various points in a medium. We must evaluate the phase differences resulting from variation of path length between wavelets. The scattered waves, as we may call these secondary waves, radiate as from a dipole source. Figure 9.2 is a schematic of the experimental

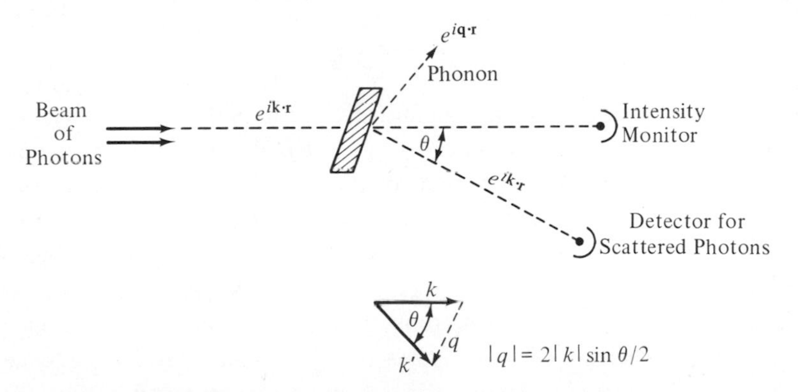

Figure 9.2 Schematic Diagram of Scattering by Beam of Photons from Atomic Target. The inset vector space diagram shows the isosceles triangle formed by the wave vectors of the incident photon, the scattered photon, and the emitted (or absorbed) phonon, which is an elastic wave in the scattering medium.

system. The scattered wave has the same frequency as the incident radiation unless the incoming photon is captured and reemitted as a photon of slightly higher or lower energy because of inelastic processes. We shall now make these remarks more quantitative.

9.3 RAYLEIGH SCATTERING

The electrons swarming around the nucleus of a simple atom such as argon in a dilute gas make a convenient example of the source of scattered photons we wish to discuss. If the incident photons are polarized, monochromatic light waves of 6800 Å, we certainly may consider the instantaneous electric vector, **E**, as constant in magnitude over the tiny diameter of the atom. Thus all the electrons of the target atom are driven by the same electric field, $\mathbf{E}e^{-i\omega t}$, although the space variation of the electric field makes for quite a different value at some distant point in the medium. The electric field at the target atom produces a dipole by pulling the electron cloud in one direction and pushing the positive nucleus in the opposite direction. The resulting dipole, p, in a static field is $p = qy = \alpha \mathbf{E}_0$ where q is the charge, y is the dipole separation, α is the polarizability, and $\mathbf{E}_0$ is the electric field. In an oscillating electric field the dipole cannot follow the driving field exactly and we may invoke the theory of Chapter 1, Eqs. 1.12 through 1.15, to describe the driven dipole as though it was a driven, damped oscillator with characteristic frequency ω_0. The dipole is simply written, according to the symbols of Eq. 1.15′ of Chapter 1, as a function of the mechanical susceptibility

$$p(t) = qy(t) = q\{q\mathbf{E}e^{-i\omega t}(\chi'(\omega) + i\chi''(\omega))\} \qquad (9.10)$$

In Eq. 1.16 of Chapter 1 we had identified the complex susceptibility, $\chi(\omega)$, with a more general response function $\chi(t - t')$. We do not have to accept the physical model of electrons held to the nucleus with a linear restoring force and a damping constant R. We can accept that an atom like argon will respond to the driving electric field of the incident radiation by producing a dipole whose time variation will certainly make possible secondary radiation. In Chapter 8 we saw that electrons bound by Coulomb force to the nucleus had characteristic energy states such that characteristic frequencies $\omega_1, \omega_2, \omega_3, \omega_j, \cdots$ are involved in the resonance response and not just some value ω_0. The dipole, $p(t)$, could then be expected to be a sum of the response contributions of each characteristic frequency with a weighting factor or "oscillator strength." The real part, neglecting damping, is

$$p(t) \simeq e^2\mathbf{E}e^{-i\omega t}\sum_j \frac{f_j}{\omega_j^2 - \omega^2} \qquad (9.11)$$

The quantity, f_j, is called the transition probability in wave mechanics or it may be called the "oscillator strength." The calculation of f_j is a quantum wave mechanical problem which is a topic for a more advanced course. Let us make use of the complex polarizability, α, to write Eq. 9.11 as

$$p(t) = \alpha(\omega)\,\mathbf{E}e^{-i\omega t} \qquad (9.12)$$

The radiation from a dipole source, according to Chapter 7, has an intensity proportional to ω^4 and to the square of the dipole strength; i.e., to $|\alpha|^2 E^2$.

For those who would like to review this classical electrodynamics, we suggest the excellent book by Arnold Sommerfeld.[4] The ratio of the total scattered radiation energy per second to the incident radiation energy per second in a beam of 1 cm^2 and at frequencies so low that α shows no dispersion, gives the effective cross section (Rayleigh 1871)

$$\sigma \simeq |\alpha|^2\omega^4 = \frac{8\pi |\alpha|^2 \omega^4 V^2}{3c^4} \tag{9.13}$$

where V is the volume of the particle and c is the velocity of light. Large atoms like argon have a much larger polarizability than a small atom like helium, and hence scatter more strongly. The experimental arrangement in Fig. 9.2 measures intensity at scattering angle θ and not the value of the electric field of the radiation. Information about the polarization of the scattered radiation could be obtained by use of Polaroid or Nicol prisms, but we have not brought out these details.

We insert into the discussion at this point a reminder which may be useful in relating our discussion with many standard texts on optics. The dipole described by Eqs. 9.10, 9.11, and 9.12 makes up a total polarization, $P = Np$ where N is the number of dipoles per unit volume. The displacement vector, $\mathbf{D}$, used in electrodynamics is related to the property of the medium and to the electric field by first a general statement $\epsilon\mathbf{E}$ and then a more specific statement; i.e., $\mathbf{D} = \epsilon\mathbf{E} = \epsilon_0\mathbf{E} + \mathbf{P}$ where ϵ is the permittivity of the medium and ϵ_0 is the permittivity of empty space. The permittivity, ϵ, is certainly a complex quantity if Eq. 9.10 is used for the dipole. In crystal optics where atoms may be arranged in special non-symmetrical arrays, the permittivity, ϵ, may be a tensor of a very special kind. We can bury our simple dipole deeper in symbolism if we define a dielectric "constant," K, through the equation $\epsilon = K\epsilon_0$. We write $K = 1 + P/\epsilon_0 E$ and attribute all the scattering mechanisms (coherent and incoherent) of a medium to the dielectric constant. If the atoms are bunched together, as in a pressure wave or an elastic wave, we do not say the number N is greater in a small volume nor that P is greater, we say the dielectric constant is greater or has fluctuated.

9.4 RAMAN AND BRILLOUIN SCATTERING

The Rayleigh scattering discussed above does not involve a change in the internal state of the molecule, and for a gas of large mean free path the

[4] A. Sommerfeld, *Electrodynamics*, page 153, Academic Press (1964); see also Sommerfeld's other books on optics and partial differential equations in physics.

scattering takes place independently at each molecule. It is possible for the incident light to interact with the molecule and change the molecule's vibration, rotation, or electronic state. The resulting scattered photon may be shifted up or down in energy (Raman scattered photon). If the quantum oscillator with which the incident photon interacts is one of the excitations of a normal mode of the elastic waves in a liquid or solid, we call the frequency shifted photon a Brillouin scattered photon. As shown in Fig. 9.2, the wave vector $\mathbf{k}$ of the photon incident and the wave vector $\mathbf{k}'$ or the scattered photon are proportional to their respective momenta, $\mathbf{k} = p/\hbar$. Since the effective momenta of the photon, phonon, photon scattering must be conserved, we have (see Fig. 9.2)

$$\mathbf{q} = |\mathbf{k}' - \mathbf{k}| = \frac{2\omega}{c} \sin \frac{\theta}{2}$$

or

$$\Delta\omega = qv_s = \pm 2\frac{\omega v_s}{c} \sin \frac{\theta}{2} \tag{9.14}$$

where v_s is the velocity of the elastic wave, c the velocity (*in situ*) of light, and ω the angular frequency of the incident photon which is taken to be the same (approximation) as that of the scattered photon. Putting in typical values, we expect a frequency shift of about $10^{-5}\nu$. The highest frequency of the elastic wave excited by the incident phonon is then of the order 10^{10} cps in a liquid or solid. These hypersonic waves are typical of the highest elastic waves making up the thermal energy in the medium. To observe these side frequencies, called Stokes lines, about the central Rayleigh line, we must use a Fabry-Perot interferometer and use monochromatic laser light in the incident beam. It is interesting to note that in a simple liquid, such as argon, the measured frequency shift[5] corresponds to using in Eq. 9.14 a velocity for the hypersonic waves which is a small departure (less than 0.4%) from that measured for the velocity of ultrasonic waves at 10^6 cps. These researchers report that there is a tentative negative dispersion of sound velocity between 10^6 and 10^{10} cps in such a simple liquid, and we would conclude that there is no resonance relaxation mechanism in this frequency range. Sound waves are attenuated in such a liquid by the energy loss caused by shear viscosity, volume viscosity, and by heat conduction. The local disturbance in the liquid relaxes exponentially to its equilibrium value with a characteristic short time, τ_i, where $i = 1, 2$, or 3 for the above energy loss mechanisms. Such mechanisms have a characteristic frequency on the pure imaginary axis of a com-

[5]P. A. Fleury and J. P. Boon, *Phys. Rev.*, **186**, 244 (1969). See also for liquid nitrogen: A. S. Pine, *J. Chem. Phys.*, **51**, 5171 (1969).

plex frequency space. No resonance frequency on the real axis would be observed in a simple liquid like argon and no normal dispersion of the velocity would occur. The line width of the Brillouin scattered photon is dependent on the above energy loss mechanisms.[6] For large complicated molecules in a liquid, Debye considered relaxation times for the reorientation, i.e., the decay of their anisotropy fluctuations, to be a slow characteristic time $\tau_D \simeq \eta a^3/kT$ where η is the shear viscosity and a is the dimension of the large molecule. There exists considerable scientific literature on the scattering of light by large molecules, and some of the best was done under Peter Debye's direction.[7] This important subject of light scattering from liquids is often discussed in special symposia by chemical physicists at meetings of the American Physical Society.

9.5 X-RAY SCATTERING AND DIFFRACTION

In the discussion of the scattering of photons by atoms and molecules in the above paragraphs we emphasized that the cluster of electrons about the atomic nucleus was driven by the incident electric field in such a way that a dipole oscillator was produced. The mechanical system of the scattering center was described by Eqs. 9.10 and 9.11, and in deriving Eq. 9.13 for simple Rayleigh scattering, the incident wave frequency was taken to be well below any resonance frequency so that the driven dipole essentially followed the oscillation of the incident electric field. Going to the other extreme, by using X-rays, the incident frequency is so much higher than the resonance frequencies that the driven oscillator is mass controlled. The mechanical impedance is dominated by the term ωm. Thus we can neglect the damping term and the linear restoring force term in the equation of motion of the electron in the high-frequency electromagnetic field of the incident X-ray, and the expression is

$$m\ddot{y} = eEe^{-i\omega t} \tag{9.15}$$

The dipole is

$$p = ey(t) = (e^2/m\omega^3)Ee^{-i\omega t}$$

and gives a scattering cross section $\sigma = 32\pi r_0^2/3$ where r_0 is the electron

[6] L. D. Landau and E. M. Lifshitz, *Electrodynamics of Continuous Media*, page 391, Addison-Wesley Publishing Co., Reading, Mass. (1966).

[7] P. Debye, *Non-Crystalline Solids*, pages 9-16, edited by V. D. Frechette, John Wiley and Sons, New York (1960).

radius (see Ref. 4). If we had an experimental situation of truly free electrons as the scattering medium in Fig. 9.2, the above equation would be true at any frequency, ω, of the incident photon and not be restricted to X-rays.

A. H. Compton (1923) showed that X-rays scattered from electrons bound in atoms, and that these electrons behaved like free electrons because they gave an elastic collision with a photon of momentum $\hbar k$. A diagram in k-space similar to that shown in Fig. 9.2 is appropriate for this electron collision.[8] The X-ray photon imparts a momentum to the electron, and the scattered photon has a new wavelength λ' and a new wave vector $\mathbf{k}'$. Modern work has measured scattered intensity as a function of scattered wave lengths: i.e., Compton line shapes. Such work gives details about the electrons in the solid.

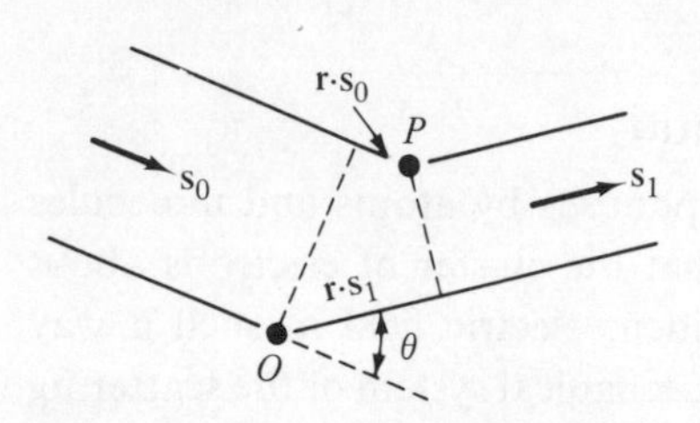

Figure 9.3 X-ray scattering from two atoms located at points O and P.

The phenomenon of diffraction results from interference effects among wave trains originating from various points in a medium. Figure 9.3 gives the scattering geometry with two scattering centers, one at 0 and the other located at a vector distance $\mathbf{r}$ at the point P. The unit vectors $\mathbf{s}_0$ and $\mathbf{s}_1$ describe the space direction of the incident and scattered wave. The difference in path length for waves scattered by these two points is just $\mathbf{r}\cdot\mathbf{s}_1 - \mathbf{r}\cdot\mathbf{s}_0$. Assuming no change in wavelength caused by the scattering centers—i.e., $k = k'$—then the corresponding difference in phase angle is $(2\pi/\lambda)\,\mathbf{r}\cdot(\mathbf{s}_1 - \mathbf{s}_0)$. Thus the wave scattered from 0 can be written $A_0e^{-i\omega t}$ and the wave scattered from P can be written $A_0e^{-i\omega t}e^{i(2\pi/\lambda)\mathbf{r}\cdot(\mathbf{s}_1-\mathbf{s}_0)}$.

The resulting wave displacement for a system of points is

$$\psi = A_0e^{-i\omega t}\sum_j e^{i(2\pi/\lambda)\mathbf{r}_j\cdot(\mathbf{s}_1-\mathbf{s}_0)} \tag{9.16}$$

This may be expressed[9] in terms of a scattering point density or electron density

$$\psi = A_0e^{-i\omega t}\int_v \rho(r)e^{i(2\pi/\lambda)\mathbf{r}\cdot(\mathbf{s}_1-\mathbf{s}_0)}\,dv \tag{9.17}$$

[8] A. H. Compton and S. K. Allison, *X-rays in Theory and Experiment*, Van Nostrand, New York (1935).

[9] See, for example, R. W. James, *The Optical Principles of the Diffraction of X-rays*, G. Bell and Sons, London (1954).

In terms of our familiar wave vector space we may write

$$\frac{2\pi}{\lambda}\mathbf{s}_0 = \mathbf{k}, \qquad \frac{2\pi}{\lambda}\mathbf{s}_1 = \mathbf{k}'$$

for the incident and scattered waves, and Eq. 9.17 becomes

$$\psi = A_0 e^{-i\omega t} \int_v \rho(r) e^{i\mathbf{q}\cdot\mathbf{r}}\, dv \tag{9.18}$$

where $\mathbf{q} = \mathbf{k} - \mathbf{k}'$ and we are *not* to consider that a phonon has been produced in this process as in Eq. 9.14 for Brillouin scattering. The corresponding intensity may be written

$$|\psi|^2 = A_0^2 \left| \int_v \rho(r) e^{i\mathbf{q}\cdot\mathbf{r}}\, dv \right|^2 \tag{9.19}$$

As $q \longrightarrow 0$ the integral in the above expression is just the electron density averaged over a unit volume—say, a lattice cell in a solid.

We have seen in Eq. 9.14 that

$$|q| = \frac{2\omega}{c} \sin\frac{\theta}{2} \tag{9.20}$$

The electron density can be described by a Fourier series in a crystal lattice with the expression

$$\rho(r) = \sum_l \rho_l e^{2\pi i \mathbf{l}\cdot\mathbf{r}} \tag{9.21}$$

where ρ_l is the electron density at reciprocal lattice point $\mathbf{l}$ and the summation is over all periods $\mathbf{l}$ of the reciprocal lattice of the crystal. Substituting in the integral expression of Eq. 9.18 we obtain

$$\psi = A_0 e^{-i\omega t} \int_v \sum_l \rho_l e^{2\pi i \mathbf{l}\cdot\mathbf{r}} \cdot e^{i\mathbf{q}\cdot\mathbf{r}}\, dv \tag{9.22}$$

The integral is practically zero except for values of $\mathbf{q}$ close to some $2\pi\mathbf{l}$; $\mathbf{q} = \mathbf{k}' - \mathbf{k} = 2\pi\mathbf{l}$ or

$$k \sin\theta/2 = \pi l \qquad \text{the Bragg diffraction maxima} \tag{9.23}$$

We may speak of Bragg reflection from crystalline planes, but the physical meaning is onc of sharp diffraction maxima. For a given $\mathbf{l}$ a principal maximum of scattering does not occur for arbitrary direction and frequency of incident radiation. Writing Eq. 9.23 as $\mathbf{k}' = \mathbf{k} + 2\pi\mathbf{l}$ and squaring this expression gives

$$k'^2 = k^2 + 4\pi\mathbf{kl} + 4\pi l^2 \tag{9.24}$$

Since (very nearly) $k^2 = k'^2$ we have that maxima occur only when the scalar product

$$\mathbf{l} \cdot \mathbf{k} = -\pi l^2 \tag{9.25}$$

Thus for a given **l** (determined by a given lattice spacing in ordinary space) we must have only the wave vector **k** to give a maximum. In Chapter 3, for electrons in a one-dimensional lattice we saw how Bragg reflection occurred when the electron wave vector **k** took on a value $1/2a$ where a was the lattice spacing.

Debye[10] showed that the periodicity of a crystal structure is not required for the production of diffraction effects. Debye and Scherrer did diffraction experiments with X-rays on liquid benzene. Debye and Menkė (1930) worked on liquid mercury. The basic problem is finding the solution to the integral in Eq. 9.19. An excellent review of the Fourier analysis of X-ray diffraction data from liquids has been given by C. J. Pings and H. H. Paalman.[11] Space does not permit us to continue this subject, but we will remark that the density function may also be a function of time; i.e., we have $\rho(r, t)$ to allow for molecular motion.

9.6 NEUTRON SCATTERING

The experiments with neutrons have taken an enormous amount of time and money during the past twenty-five years since the availability of a huge flux of slow, monochromatic neutrons from modern nuclear reactors. The scattering target for the neutron beam is the nucleus itself and the potential, $V(r)$, exists because of a magnetic moment of both particles. Inside the nucleus the potential is something else. Figure 9.4 gives a schematic of the modern neutron scattering experiment which is capable of measuring the energy and energy changes of scattered neutrons. For X-rays, due to the extremely small relative size of the energy changes, the dispersion law has to be inferred indirectly from intensity measurements. But for slow neutrons the energy change can be large and the knowledge gained of crystal vibrations and of molecular motions is most important in inelastic scattering of neu-

[10] P. Debye, *Ann. Physik*, **46**, 809 (1915); *Zeit. Physik*, **28**, 135 (1927).
[11] H. H. Paalman and C. J. Pings, *Rev. Mod. Phys.*, **35**, 389 (1963).

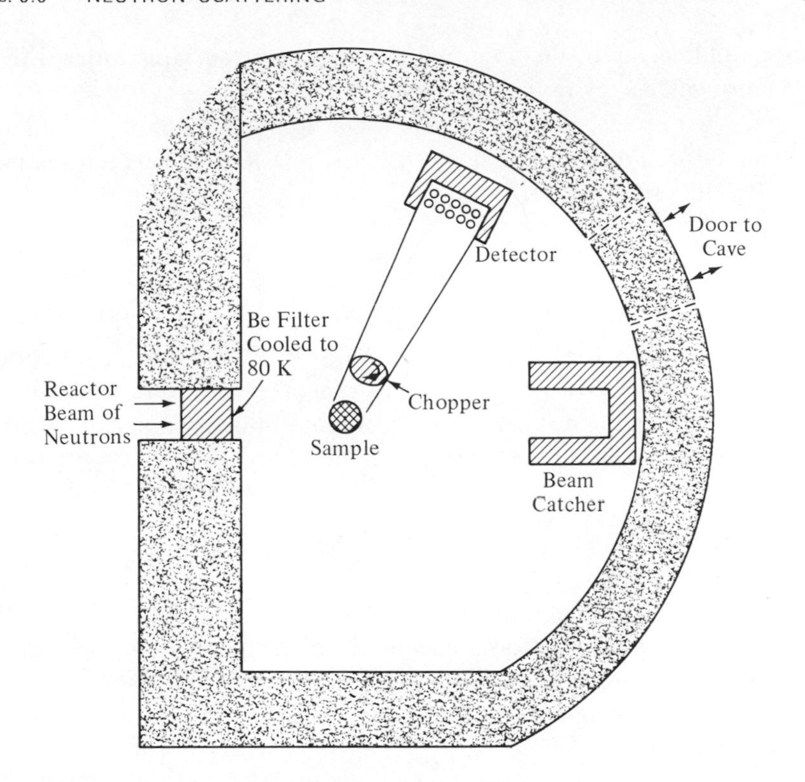

Figure 9.4 Neutron Scattering Experimental Cave attached to beam port of nuclear reactor. The Neutron Spectrometer (time of flight technique) is used in studies of atomic and molecular motion.

trons.[12] The diffraction of neutrons by elastic scattering in crystals is in full analogy with the Bragg reflection of X-rays (coherent radiation), and we shall not repeat this part of the subject.[13]

Suppose the incident neutron beam were to be so "cold" that the wavelength approaches $\lambda \longrightarrow \infty$. Then a single phonon in the crystal target strikes the neutron and gives it all the energy the scattered neutron has

$$\frac{k^2\hbar^2}{2m} = \hbar\omega_j(\mathbf{q}) \tag{9.26}$$

where m and $\mathbf{k}$ are the mass and final wave vector of the neutron and $\mathbf{q}$, j, and $\hbar\,\omega_j(\mathbf{q})$ are the wave vector, polarization index ($j = 1, 2, 3$ for a Bravais

[12]G. Placzek and L. Van Hove, *Phys. Rev.*, **93**, 1207 (1953); L. Van Hove, *Phys. Rev.*, **95**, 249 (1954).

[13]See, for example, D. J. Hughes, *Pile Neutron Research*, Addison-Wesley Publishing Co., Reading, Mass. (1953); also P. A. Egelstaff, *Thermal Neutron Scattering*, Academic Press, New York (1965).

lattice), and energy of the absorbed phonon. Neutrons are scattered in all directions with energy ranging up to $\hbar\omega_{\max}$.

We may speak of the neutron momentum $\hbar\mathbf{k}$ and of the phonon momentum $\hbar\mathbf{q}$. For coherent scattering, the interference between scattered waves by the various nuclei requires

$$\mathbf{k} = \mathbf{q} + 2\pi\mathbf{l} \tag{9.27}$$

where $\mathbf{l}$ is an arbitrary vector of the reciprocal lattice. We recall that the lattice wave vector repeats itself each $2\pi l$; i.e., $e^{i\mathbf{q}x} = e^{i(\mathbf{q}+2\pi\mathbf{l})x}$ owing to the periodicity of the lattice. Also we have $\omega_j(\mathbf{q}) = \omega_j(\mathbf{q} + 2\pi\mathbf{l})$ so that Eq. 9.27 is conventionally regarded as conservation of momentum. For coherent scattering the restriction Eq. 9.27 requires

$$k^2 = \frac{2m}{\hbar}\,\omega_j(\mathbf{k}) \tag{9.28}$$

which defines a surface in k space for each value $j = 1, 2, 3$. Placzek and Van Hove[12] pointed out that measurements of outgoing energy of the neutrons as a function of direction, determines the surface and thus yields $\omega_j(\mathbf{k})$. This type of scattering is particularly well fitted to give information on the crystal vibrations.

Table of Physical Constants* and Useful Values

MKS units where J = joules, K = degrees Kelvin, N = newtons, T = webers/m^2

Speed of light in vacuum, c	2.997925×10^8 m/sec
Electronic charge, e	1.60210×10^{-19} C
Electron rest mass, m_e	9.1091×10^{-31} kg
Charge to mass ratio for electron, e/m_e	1.758796×10^{11} C/kg
Proton rest mass, m_p	1.67252×10^{-27} kg
Ratio of mass of a proton to that of an electron, m_p/m_e	1,837
Avogadro constant, N_A	6.02252×10^{23} mol^{-1}
Faraday constant, F	9.64870×10^4 C/mol
Planck constant, h	6.6256×10^{-34} J-sec
Rydberg constant, R_∞	1.0973731×10^7 m^{-1}
Gas constant, R	8.3143 J/K mol
Boltzmann constant, k	1.38054×10^{-23} J/K
Gravitational constant, G	6.670×10^{-11} N m^2/kg^2
Bohr magneton, μ_B	9.2732×10^{-24} J/T
Permittivity of free space, ϵ_0	8.854×10^{-12} farad/m
Permeability of free space, μ_0	$4\pi \times 10^{-7}$ henry/m
Gyromagnetic ratio of proton, γ	2.67519×10^8 rad/sec T
Energy units: 1 joule = 10^7 ergs,	1 calorie = 4.19×10^7 ergs

Speed of sound	in air	density 0.00129 gm/cm^3	0°C & 1 Atm.	331.00 m/sec
	in pure water	1.000 gm/cm^3	4°C	1430.0 m/sec
	in sea water	1.023 gm/cm^3	4°C	1447.0 m/sec
(Longi.)	in solid steel	7.90 gm/cm^3	20°C	5850 m/sec
(Longi.)	in solid beryllium	1.82 gm/cm^3	300 K	12,550 m/sec

*From National Bureau of Standards, *News Bull.*, Oct. 1963.

INDEX